HIGHLIGHTS OF
Climate Change Impacts in the United States

Observed U.S. Temperature Change

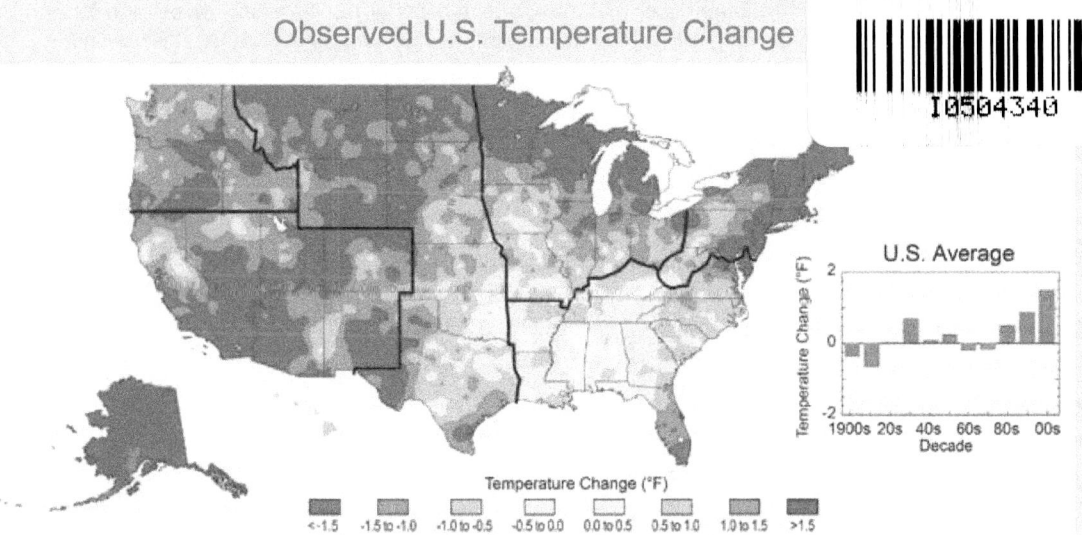

The colors on the map show temperature changes over the past 22 years (1991-2012) compared to the 1901-1960 average for the contiguous U.S., and to the 1951-1980 average for Alaska and Hawai'i. The bars on the graph show the average temperature changes for the U.S. by decade for 1901-2012 (relative to the 1901-1960 average). The far right bar (2000s decade) includes 2011 and 2012. The period from 2001 to 2012 was warmer than any previous decade in every region. (Figure source: NOAA NCDC / CICS-NC).

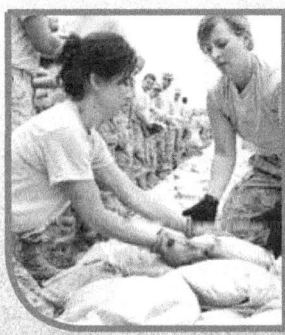

Members of the National Guard lay sandbags to protect against Missouri River flooding.

Energy choices will affect the amount of future climate change.

Climate change is increasing the vulnerability of forests to wildfires across the U.S. West.

Solar power use is increasing and is part of the solution to climate change.

Online at:
nca2014.globalchange.gov

Recommended Citation

Printed and Published by U.S. Government Printing Office
Internet: bookstore.gpo.gov; Phone: toll free (866) 512-1800; DC area (202) 512-1800
Fax: (202) 512-2104 Mail: Stop IDCC, Washington, DC 20402-0001

May 2014

Members of Congress:

On behalf of the National Science and Technology Council and the U.S. Global Change Research Program, we are pleased to transmit the report of the Third National Climate Assessment: *Climate Change Impacts in the United States.* As required by the Global Change Research Act of 1990, this report has collected, evaluated, and integrated observations and research on climate change in the United States. It focuses both on changes that are happening now and further changes that we can expect to see throughout this century.

This report is the result of a three-year analytical effort by a team of over 300 experts, overseen by a broadly constituted Federal Advisory Committee of 60 members. It was developed from information and analyses gathered in over 70 workshops and listening sessions held across the country. It was subjected to extensive review by the public and by scientific experts in and out of government, including a special panel of the National Research Council of the National Academy of Sciences. This process of unprecedented rigor and transparency was undertaken so that the findings of the National Climate Assessment would rest on the firmest possible base of expert judgment.

We gratefully acknowledge the authors, reviewers, and staff who have helped prepare this Third National Climate Assessment. Their work in assessing the rapid advances in our knowledge of climate science over the past several years has been outstanding. Their findings and key messages not only describe the current state of that science but also the current and future impacts of climate change on major U.S. regions and key sectors of the U.S. economy. This information establishes a strong base that government at all levels of U.S. society can use in responding to the twin challenges of changing our policies to mitigate further climate change and preparing for the consequences of the climate changes that can no longer be avoided. It is also an important scientific resource to empower communities, businesses, citizens, and decision makers with information they need to prepare for and build resilience to the impacts of climate change.

When President Obama launched his Climate Action Plan last year, he made clear that the essential information contained in this report would be used by the Executive Branch to underpin future policies and decisions to better understand and manage the risks of climate change. We strongly and respectfully urge others to do the same.

Sincerely,

Dr. John P. Holdren
Assistant to the President for Science and Technology
Director, Office of Science and Technology Policy
Executive Office of the President

Dr. Kathryn D. Sullivan
Under Secretary for Oceans and Atmosphere
NOAA Administrator
U.S. Department of Commerce

Report Authors and Additional Staff, see page 98

The National Climate Assessment assesses the science of climate change and its impacts across the United States, now and throughout this century. It documents climate change related impacts and responses for various sectors and regions, with the goal of better informing public and private decision-making at all levels.

A team of more than 300 experts (see page 98), guided by a 60-member National Climate Assessment and Development Advisory Committee (listed on page ii) produced the full report – the largest and most diverse team to produce a U.S. climate assessment. Stakeholders involved in the development of the assessment included decision-makers from the public and private sectors, resource and environmental managers, researchers, representatives from businesses and non-governmental organizations, and the general public. More than 70 workshops and listening sessions were held, and thousands of public and expert comments on the draft report provided additional input to the process.

The assessment draws from a large body of scientific peer-reviewed research, technical input reports, and other publicly available sources; all sources meet the standards of the Information Quality Act. The report was extensively reviewed by the public and experts, including a panel of the National Academy of Sciences, the 13 Federal agencies of the U.S. Global Change Research Program, and the Federal Committee on Environment, Natural Resources, and Sustainability.

Online at:
nca2014.globalchange.gov

This book presents the major findings and selected highlights from *Climate Change Impacts in the United States*, the third National Climate Assessment.

This *Highlights* report is organized around the National Climate Assessment's 12 Report Findings, which take an overarching view of the entire report and its 30 chapters. All material in the *Highlights* report is drawn from the full report. The Key Messages from each of the 30 report chapters appear in boxes throughout this document.

In the lower left corner of each section, icons identify which chapters of the full report were drawn upon for that section. A key to these icons appears on page 1.

A 20-page *Overview* booklet is available online.

Online at:
nca2014.globalchange.gov/highlights

CONTENTS

REPORT FINDINGS

REGIONS

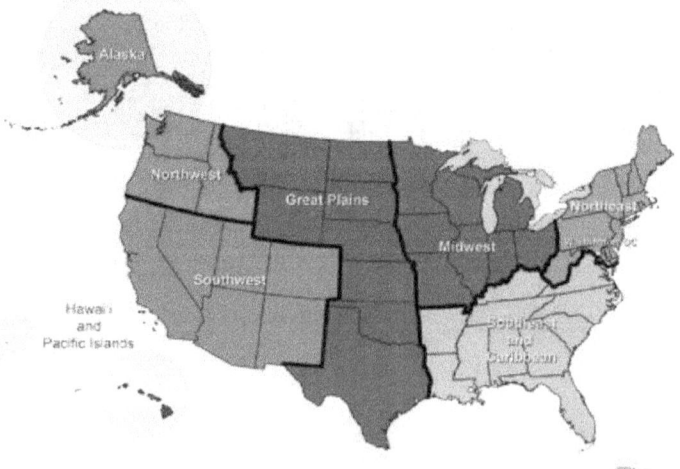

CHAPTER ICONS

In the lower left corner of each section, these icons identify which chapters of the full report were drawn upon for that section.

 Our Changing Climate

 Water Resources

 Energy Supply and Use

 Transportation

 Agriculture

 Forests

 Ecosystems and Biodiversity

 Human Health

 Energy, Water, and Land Use

 Urban Systems and Infrastructure

 Indigenous Peoples, Lands, and Resources

 Land Use and Land Cover Change

 Rural Communities

 Biogeochemical Cycles

 Northeast

 Southeast and Caribbean

 Midwest

 Great Plains

 Southwest

 Northwest

 Alaska

 Hawai'i and U.S. Affiliated Pacific Islands

 Oceans and Marine Resources

 Coastal Zones

 Decision Support

 Mitigation

 Adaptation

Appendix 3: Climate Science Supplement

 Appendix 4: Frequently Asked Questions

Climate change, once considered an issue for a distant future, has moved firmly into the present. Corn producers in Iowa, oyster growers in Washington State, and maple syrup producers in Vermont are all observing climate-related changes that are outside of recent experience. So, too, are coastal planners in Florida, water managers in the arid Southwest, city dwellers from Phoenix to New York, and Native Peoples on tribal lands from Louisiana to Alaska. This National Climate Assessment concludes that the evidence of human-induced climate change continues to strengthen and that impacts are increasing across the country.

Americans are noticing changes all around them. Summers are longer and hotter, and extended periods of unusual heat last longer than any living American has ever experienced. Winters are generally shorter and warmer. Rain comes in heavier downpours. People are seeing changes in the length and severity of seasonal allergies, the plant varieties that thrive in their gardens, and the kinds of birds they see in any particular month in their neighborhoods.

Other changes are even more dramatic. Residents of some coastal cities see their streets flood more regularly during storms and high tides. Inland cities near large rivers also experience more flooding, especially in the Midwest and Northeast. Insurance rates are rising in some vulnerable locations, and insurance is no longer available in others. Hotter and drier weather and earlier snow melt mean that wildfires in the West start earlier in the spring, last later into the fall, and burn more acreage. In Arctic Alaska, the summer sea ice that once protected the coasts has receded, and autumn storms now cause more erosion, threatening many communities with relocation.

Scientists who study climate change confirm that these observations are consistent with significant changes in Earth's climatic trends. Long-term, independent records from weather stations, satellites, ocean buoys, tide gauges, and many other data sources all confirm that our nation, like the rest of the world, is warming. Precipitation patterns are changing, sea level is rising, the oceans are becoming more acidic, and the frequency and intensity of some extreme weather events are increasing. Many lines of independent evidence demonstrate that the rapid warming of the past half-century is due primarily to human activities.

The observed warming and other climatic changes are triggering wide-ranging impacts in every region of our country and throughout our economy. Some of these changes can be beneficial over the short run, such as a longer growing season in some regions and a longer shipping season on

the Great Lakes. But many more are detrimental, largely because our society and its infrastructure were designed for the climate that we have had, not the rapidly changing climate we now have and can expect in the future. In addition, climate change does not occur in isolation. Rather, it is superimposed on other stresses, which combine to create new challenges.

This National Climate Assessment collects, integrates, and assesses observations and research from around the country, helping us to see what is actually happening and understand what it means for our lives, our livelihoods, and our future. The report includes analyses of impacts on seven sectors – human health, water, energy, transportation, agriculture, forests, and ecosystems – and the interactions among sectors at the national level. The report also assesses key impacts on all U.S. regions: Northeast, Southeast and Caribbean, Midwest, Great Plains, Southwest, Northwest, Alaska, Hawai`i and Pacific Islands, as well as the country's coastal areas, oceans, and marine resources.

Over recent decades, climate science has advanced significantly. Increased scrutiny has led to increased certainty that we are now seeing impacts associated with human-induced climate change. With each passing year, the accumulating evidence further expands our understanding and extends the record of observed trends in temperature, precipitation, sea level, ice mass, and many other variables recorded by a variety of measuring systems and analyzed by independent research groups from around the world. It is notable that as these data records have grown longer and climate models have become more comprehensive, earlier predictions have largely been confirmed. The only real surprises have been that some changes, such as sea level rise and Arctic sea ice decline, have outpaced earlier projections.

What is new over the last decade is that we know with increasing certainty that climate change is happening now. While scientists continue to refine projections of the future, observations unequivocally show that climate is changing and that the warming of the past 50 years is primarily due to human-induced emissions of heat-trapping gases. These emissions come mainly from burning coal, oil, and gas, with additional contributions from forest clearing and some agricultural practices.

Global climate is projected to continue to change over this century and beyond, but there is still time to act to limit the amount of change and the extent of damaging impacts.

This report documents the changes already observed and those projected for the future.

It is important that these findings and response options be shared broadly to inform citizens and communities across our nation. Climate change presents a major challenge for society. This report advances our understanding of that challenge and the need for the American people to prepare for and respond to its far-reaching implications.

OVERVIEW

Climate change is already affecting the American people in far-reaching ways. Certain types of extreme weather events with links to climate change have become more frequent and/or intense, including prolonged periods of heat, heavy downpours, and, in some regions, floods and droughts. In addition, warming is causing sea level to rise and glaciers and Arctic sea ice to melt, and oceans are becoming more acidic as they absorb carbon dioxide. These and other aspects of climate change are disrupting people's lives and damaging some sectors of our economy.

Coal-fired power plants emit heat-trapping carbon dioxide to the atmosphere.

Climate Change: Present and Future

Evidence for climate change abounds, from the top of the atmosphere to the depths of the oceans. Scientists and engineers from around the world have meticulously collected this evidence, using satellites and networks of weather balloons, thermometers, buoys, and other observing systems. Evidence of climate change is also visible in the observed and measured changes in location and behavior of species and functioning of ecosystems. Taken together, this evidence tells an unambiguous story: the planet is warming, and over the last half century, this warming has been driven primarily by human activity.

Multiple lines of independent evidence confirm that human activities are the primary cause of the global warming of the past 50 years. The burning of coal, oil, and gas, and clearing of forests have increased the concentration of carbon dioxide in the atmosphere by more than 40% since the Industrial Revolution, and it has been known for almost two centuries that this carbon dioxide traps heat. Methane and nitrous oxide emissions from agriculture and other human activities add to the atmospheric burden of heat-trapping gases. Data show that natural factors like the sun and volcanoes cannot have caused the warming observed over the past 50 years. Sensors on satellites have measured the sun's output with great accuracy and found no overall increase during the past half century. Large volcanic eruptions during this period, such as Mount Pinatubo in 1991, have exerted a short-term *cooling* influence. In fact, if not for human activities, global climate would actually have cooled slightly over the past 50 years. The pattern of temperature change through the layers of the atmosphere, with warming near the surface and cooling higher up in the stratosphere, further confirms that it is the buildup of heat-trapping gases (also known as "greenhouse gases") that has caused most of the Earth's warming over the past half century.

Because human-induced warming is superimposed on a

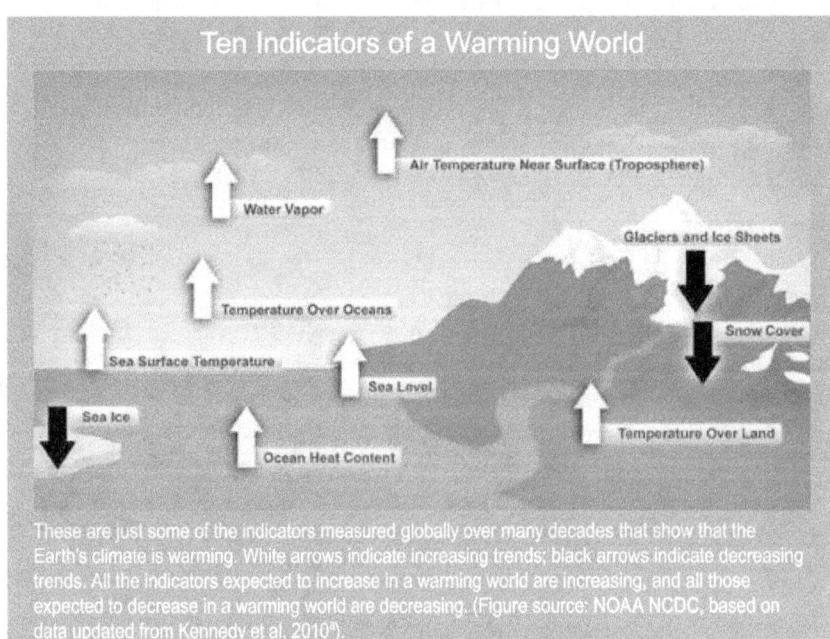

Ten Indicators of a Warming World

Air Temperature Near Surface (Troposphere)

Water Vapor

Glaciers and Ice Sheets

Temperature Over Oceans

Snow Cover

Sea Surface Temperature

Sea Level

Sea Ice

Temperature Over Land

Ocean Heat Content

These are just some of the indicators measured globally over many decades that show that the Earth's climate is warming. White arrows indicate increasing trends; black arrows indicate decreasing trends. All the indicators expected to increase in a warming world are increasing, and all those expected to decrease in a warming world are decreasing. (Figure source: NOAA NCDC, based on data updated from Kennedy et al. 2010[8]).

background of natural variations in climate, warming is not uniform over time. Short-term fluctuations in the long-term upward trend are thus natural and expected. For example, a recent slowing in the rate of surface air temperature rise appears to be related to cyclic changes in the oceans and in the sun's energy output, as well as a series of small volcanic eruptions and other factors. Nonetheless, global temperatures are still on the rise and are expected to rise further.

U.S. average temperature has increased by 1.3°F to 1.9°F since 1895, and most of this increase has occurred since 1970. The most recent decade was the nation's and the world's hottest on record, and 2012 was the hottest year on record in the continental United States. All U.S. regions have experienced warming in recent decades, but the extent of warming has not been uniform. In general, temperatures are rising more quickly in the north. Alaskans have experienced some of the largest increases in temperature between 1970 and the present. People living in the Southeast have experienced some of the smallest temperature increases over this period.

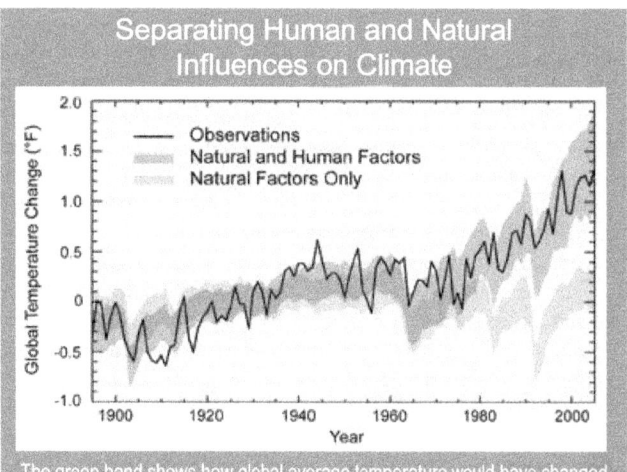

Separating Human and Natural Influences on Climate

The green band shows how global average temperature would have changed over the last century due to natural forces alone, as simulated by climate models. The blue band shows model simulations of the effects of human and natural forces (including solar and volcanic activity) combined. The black line shows the actual observed global average temperatures. Only with the inclusion of human influences can models reproduce the observed temperature changes. (Figure source: adapted from Huber and Knutti 2012[b]).

Temperatures are projected to rise another 2°F to 4°F in most areas of the United States over the next few decades. Reductions in some short-lived human-induced emissions that contribute to warming, such as black carbon (soot) and methane, could reduce some of the projected warming over the next couple of decades, because, unlike carbon dioxide, these gases and particles have relatively short atmospheric lifetimes.

The amount of warming projected beyond the next few decades is directly linked to the cumulative global emissions of heat-trapping gases and particles. By the end of this century, a roughly 3°F to 5°F rise is projected under a lower emissions scenario, which would require substantial reductions in emissions (referred to as the "B1 scenario"), and a 5°F to 10°F rise for a higher emissions scenario assuming continued increases in emissions, predominantly from fossil fuel combustion (referred to as the "A2 scenario"). These projections are based on results from 16 climate models that used the two emissions scenarios in a formal inter-model comparison study. The range of model projections for each emissions scenario is the result of the differences in the ways the models represent key factors such as water vapor, ice and snow reflectivity, and clouds, which can either dampen or amplify the initial effect of human influences on temperature. The net effect of these feedbacks is expected to amplify warming. More information about the models and scenarios used in this report can be found in Appendix 5 of the full report.[1]

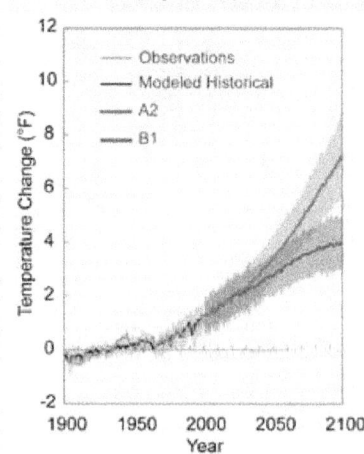

Projected Global Temperature Change

Different amounts of heat-trapping gases released into the atmosphere by human activities produce different projected increases in Earth's temperature. The lines on the graph represent a central estimate of global average temperature rise (relative to the 1901-1960 average) for the two main scenarios used in this report. A2 assumes continued increases in emissions throughout this century, and B1 assumes significant emissions reductions, though not due explicitly to climate change policies. Shading indicates the range (5th to 95th percentile) of results from a suite of climate models. In both cases, temperatures are expected to rise, although the difference between lower and higher emissions pathways is substantial. (Figure source: NOAA NCDC / CICS-NC).

Prolonged periods of high temperatures and the persistence of high nighttime temperatures have increased in many locations (especially in urban areas) over the past half century. High nighttime temperatures have widespread impacts because people, livestock, and wildlife get no respite from the heat. In some regions, prolonged periods of high temperatures associated with droughts contribute to conditions that lead to larger wildfires and longer fire seasons. As expected in a warming climate, recent trends show that extreme heat is becoming more common, while extreme cold is becoming less common. Evidence indicates that the human influence on climate has already roughly doubled the probability of extreme heat events such as the record-breaking summer heat experienced in 2011 in Texas and Oklahoma. The incidence of record-breaking high temperatures is projected to rise.[2]

Human-induced climate change means much more than just hotter weather. Increases in ocean and freshwater temperatures, frost-free days, and heavy downpours have all been documented. Global sea level has risen, and there have been large reductions in snow-cover extent, glaciers, and sea ice. These changes and other climatic changes have affected and will continue to affect human health, water supply, agriculture, transportation, energy, coastal areas, and many other sectors of society, with increasingly adverse impacts on the American economy and quality of life.[3]

Some of the changes discussed in this report are common to many regions. For example, large increases in heavy precipitation have occurred in the Northeast, Midwest, and Great Plains, where heavy downpours have frequently led to runoff that exceeded the capacity of storm drains and levees, and caused flooding events and accelerated erosion. Other impacts, such as those associated with the rapid thawing of permafrost in Alaska, are unique to a particular U.S. region. Permafrost thawing is causing extensive damage to infrastructure in our nation's largest state.[4]

Some impacts that occur in one region ripple beyond that region. For example, the dramatic decline of summer sea ice in the Arctic – a loss of ice cover roughly equal to half the area of the continental United States – exacerbates global warming by reducing the reflectivity of Earth's surface and increasing the amount of heat absorbed. Similarly, smoke from wildfires in one location can contribute to poor air quality in faraway regions, and evidence suggests that particulate matter can affect atmospheric properties and therefore weather patterns. Major storms and the higher storm surges exacerbated by sea level rise that hit the Gulf Coast affect the entire country through their cascading effects on oil and gas production and distribution.[5]

Water expands as it warms, causing global sea levels to rise; melting of land-based ice also raises sea level by adding water to the oceans. Over the past century, global average sea level has risen by about 8 inches. Since 1992, the rate of global sea level rise measured by satellites has been roughly twice the rate observed over the last century, providing evidence of acceleration. Sea level rise,

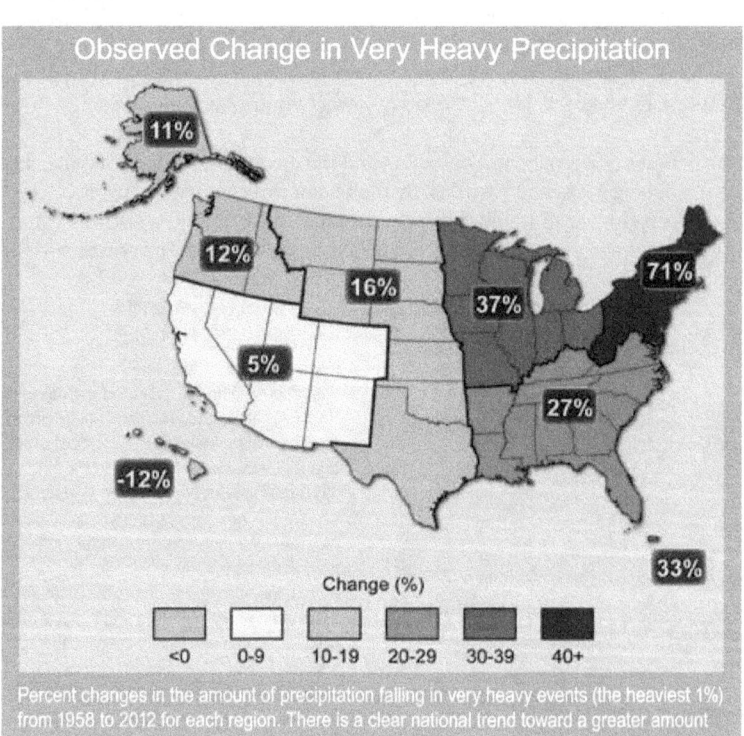

Observed Change in Very Heavy Precipitation

11%
12%
16%
37%
71%
5%
27%
-12%
33%

Change (%)

| <0 | 0-9 | 10-19 | 20-29 | 30-39 | 40+ |

Percent changes in the amount of precipitation falling in very heavy events (the heaviest 1%) from 1958 to 2012 for each region. There is a clear national trend toward a greater amount of precipitation being concentrated in very heavy events, particularly in the Northeast and Midwest. (Figure source: updated from Karl et al. 2009[6]).

combined with coastal storms, has increased the risk of erosion, storm surge damage, and flooding for coastal communities, especially along the Gulf Coast, the Atlantic seaboard, and in Alaska. Coastal infrastructure, including roads, rail lines, energy infrastructure, airports, port facilities, and military bases, are increasingly at risk from sea level rise and damaging storm surges. Sea level is projected to rise by another 1 to 4 feet in this century, although the rise in sea level in specific regions is expected to vary from this global average for a number of reasons. A wider range of scenarios, from 8 inches to more than 6 feet by 2100, has been used in risk-based analyses in this report. In general, higher emissions scenarios that lead to more warming would be expected to lead to higher amounts of sea level rise. The stakes are high, as nearly five million Americans and hundreds of billions of dollars of property are located in areas that are less than four feet above the local high-tide level.[6]

In addition to causing changes in climate, increasing levels of carbon dioxide from the burning of fossil fuels and other human activities have a direct effect on the world's

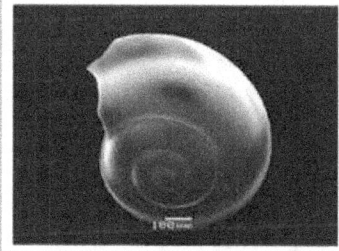

Shells Dissolve in Acidified Ocean Water

Pteropods, or "sea butterflies," are eaten by a variety of marine species ranging from tiny krill to salmon to whales. The photos show what happens to a pteropod's shell in seawater that is too acidic. On the left is a shell from a live pteropod from a region in the Southern Ocean where acidity is not too high. The shell on the right is from a pteropod in a region where the water is more acidic. (Figure source: (left) Bednaršek et al. 2012[8] (right) Nina Bednaršek).

oceans. Carbon dioxide interacts with ocean water to form carbonic acid, increasing the ocean's acidity. Ocean surface waters have become 30% more acidic over the last 250 years as they have absorbed large amounts of carbon dioxide from the atmosphere. This ocean acidification makes water more corrosive, reducing the capacity of marine organisms with shells or skeletons made of calcium carbonate (such as corals, krill, oysters, clams, and crabs) to survive, grow, and reproduce, which in turn will affect the marine food chain.[7]

Widespread Impacts

Impacts related to climate change are already evident in many regions and sectors and are expected to become increasingly disruptive across the nation throughout this century and beyond. Climate changes interact with other environmental and societal factors in ways that can either moderate or intensify these impacts.

Some climate changes currently have beneficial effects for specific sectors or regions. For example, current benefits of warming include longer growing seasons for agriculture and longer ice-free periods for shipping on the Great Lakes. At the same time, however, longer growing seasons, along with higher temperatures and carbon dioxide levels, can increase pollen production, intensifying and lengthening the allergy season. Longer ice-free periods on the Great Lakes can result in more lake-effect snowfalls.

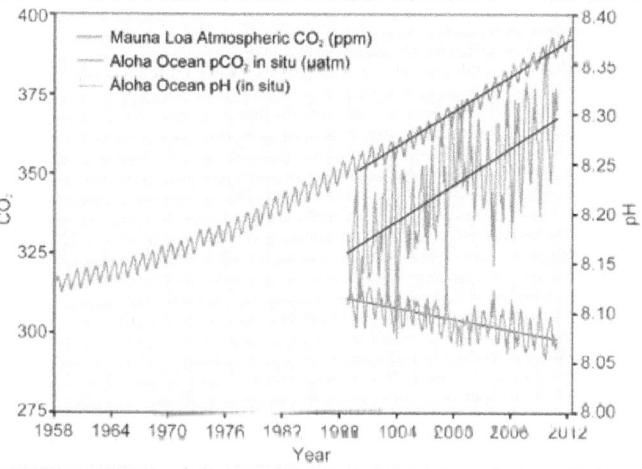

As Oceans Absorb CO$_2$ They Become More Acidic

The correlation between rising levels of carbon dioxide in the atmosphere (red) with rising carbon dioxide levels (blue) and falling pH in the ocean (green). As carbon dioxide accumulates in the ocean, the water becomes more acidic (the pH declines). (Figure source: modified from Feely et al. 2009[d]).

Observed and projected climate change impacts vary across the regions of the United States. Selected impacts emphasized in the regional chapters are shown below, and many more are explored in detail in this report.

	Region	Impact
	Northeast	Communities are affected by heat waves, more extreme precipitation events, and coastal flooding due to sea level rise and storm surge.
	Southeast and Caribbean	Decreased water availability, exacerbated by population growth and land-use change, causes increased competition for water. There are increased risks associated with extreme events such as hurricanes.
	Midwest	Longer growing seasons and rising carbon dioxide levels increase yields of some crops, although these benefits have already been offset in some instances by occurrence of extreme events such as heat waves, droughts, and floods.
	Great Plains	Rising temperatures lead to increased demand for water and energy and impacts on agricultural practices.
	Southwest	Drought and increased warming foster wildfires and increased competition for scarce water resources for people and ecosystems.
	Northwest	Changes in the timing of streamflow related to earlier snowmelt reduce the supply of water in summer, causing far-reaching ecological and socioeconomic consequences.
	Alaska	Rapidly receding summer sea ice, shrinking glaciers, and thawing permafrost cause damage to infrastructure and major changes to ecosystems. Impacts to Alaska Native communities increase.
	Hawai'i and Pacific Islands	Increasingly constrained freshwater supplies, coupled with increased temperatures, stress both people and ecosystems and decrease food and water security.
	Coasts	Coastal lifelines, such as water supply infrastructure and evacuation routes, are increasingly vulnerable to higher sea levels and storm surges, inland flooding, and other climate-related changes.
	Oceans	The oceans are currently absorbing about a quarter of human-caused carbon dioxide emissions to the atmosphere and over 90% of the heat associated with global warming, leading to ocean acidification and the alteration of marine ecosystems.

Sectors affected by climate changes include agriculture, water, human health, energy, transportation, forests, and ecosystems. Climate change poses a major challenge to U.S. agriculture because of the critical dependence of agricultural systems on climate. Climate change has the potential to both positively and negatively affect the location, timing, and productivity of crop, livestock, and fishery systems at local, national, and global scales. The United States produces nearly $330 billion per year in agricultural commodities. This productivity is vulnerable to direct impacts on crops and livestock from changing climate conditions and extreme weather events and indirect impacts through increasing pressures from pests and pathogens. Climate change will also alter the stability of food supplies and create new food security challenges for the United States as the world seeks to feed nine billion people by 2050. While the agriculture sector has proven to be adaptable to a range of stresses, as evidenced by continued growth in production and efficiency across the United States, climate change poses a new set of challenges.[8]

Water quality and quantity are being affected by climate change. Changes in precipitation and runoff, combined with changes in consumption and withdrawal, have reduced surface and groundwater supplies in many areas. These trends are expected to continue, increasing the likelihood of water shortages for many uses. Water quality is also diminishing in many areas, particularly due to sediment and contaminant concentrations after heavy downpours.

Increasing air and water temperatures, more intense precipitation and runoff, and intensifying droughts can decrease water quality in many ways. Here, middle school students in Colorado test water quality.

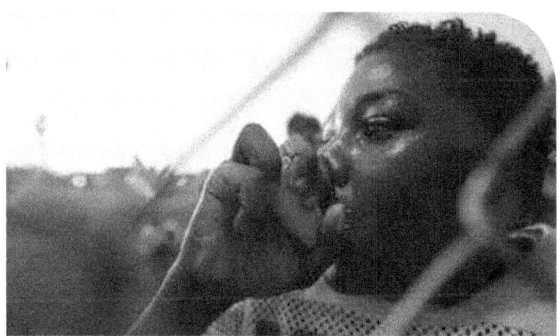

Climate change can exacerbate respiratory and asthma-related conditions through increases in pollen, ground-level ozone, and wildfire smoke.

Sea level rise, storms and storm surges, and changes in surface and groundwater use patterns are expected to compromise the sustainability of coastal freshwater aquifers and wetlands. In most U.S. regions, water resources managers and planners will encounter new risks, vulnerabilities, and opportunities that may not be properly managed with existing practices.[9]

Certain groups of people are more vulnerable to the range of climate change related health impacts, including the elderly, children, the poor, and the sick.

Climate change affects human health in many ways. For example, increasingly frequent and intense heat events lead to more heat-related illnesses and deaths and, over time, worsen drought and wildfire risks, and intensify air pollution. Increasingly frequent extreme precipitation and associated flooding can lead to injuries and increases in waterborne disease. Rising sea surface temperatures have been linked with increasing levels and ranges of diseases. Rising sea levels intensify coastal flooding and storm surge, and thus exacerbate threats to public safety during storms. Certain groups of people are more vulnerable to the range of climate change related health impacts, including the elderly, children, the poor, and the sick. Others are vulnerable because of where they live, including those in floodplains, coastal zones, and some urban areas. Improving and properly supporting the public health infrastructure will be critical to managing the potential health impacts of climate change.[10]

Climate change also affects the living world, including people, through changes in ecosystems and biodiversity. Ecosystems provide a rich array of benefits and services to humanity, including habitat for fish and wildlife, drinking water storage and filtration, fertile soils for growing crops, buffering against a range of stressors including climate change impacts, and aesthetic and cultural values. These benefits are not always easy to quantify, but they support jobs, economic growth, health, and human well-being. Climate change driven disruptions to ecosystems have direct and indirect human impacts, including reduced water supply and quality, the loss of iconic species and landscapes, effects on food chains and the timing and success of species migrations, and the potential for extreme weather and climate events to destroy or degrade the ability of ecosystems to provide societal benefits.[11]

Human modifications of ecosystems and landscapes often increase their vulnerability to damage from extreme weather events, while simultaneously reducing their natural capacity to moderate the impacts of such events. For example, salt marshes, reefs, mangrove forests, and barrier islands defend coastal ecosystems and infrastructure, such as roads and buildings, against storm surges. The loss of these natural buffers due to coastal development, erosion, and sea level rise increases the risk of catastrophic damage during or after extreme weather events. Although floodplain wetlands are greatly reduced from their historical extent, those that remain still absorb floodwaters and reduce the effects of high flows on river-margin lands. Extreme weather events that produce sudden increases in water flow, often carrying debris and pollutants, can decrease the natural capacity of ecosystems to cleanse contaminants.[12]

> The amount of future climate change will still largely be determined by choices society makes about emissions.

The climate change impacts being felt in the regions and sectors of the United States are affected by global trends and economic decisions. In an increasingly interconnected world, U.S. vulnerability is linked to impacts in other nations. It is thus difficult to fully evaluate the impacts of climate change on the United States without considering consequences of climate change elsewhere.

Response Options

As the impacts of climate change are becoming more prevalent, Americans face choices. Especially because of past emissions of long-lived heat-trapping gases, some additional climate change and related impacts are now unavoidable. This is due to the long-lived nature of many of these gases, as well as the amount of heat absorbed and retained by the oceans and other responses within the climate system. The amount of future climate change, however, will still largely be determined by choices society makes about emissions. Lower emissions of heat-trapping

gases and particles mean less future warming and less-severe impacts; higher emissions mean more warming and more severe impacts. Efforts to limit emissions or increase carbon uptake fall into a category of response options known as "mitigation," which refers to reducing the amount and speed of future climate change by reducing emissions of heat-trapping gases or removing carbon dioxide from the atmosphere.[13]

The other major category of response options is known as "adaptation," and refers to actions to prepare for and adjust to new conditions, thereby reducing harm or taking advantage of new opportunities. Mitigation and adaptation actions are linked in multiple ways, including that effective mitigation reduces the need for adaptation in the future. Both are essential parts of a comprehensive climate change response strategy. The threat of irreversible impacts makes the timing of mitigation efforts particularly critical. This report includes chapters on Mitigation, Adaptation, and Decision Support that offer an overview of the options and activities being planned or implemented around the country as local, state, federal, and tribal governments, as well as businesses, organizations, and individuals begin to respond to climate change. These chapters conclude that while response actions are under development, current implementation efforts are insufficient to avoid increasingly negative social, environmental, and economic consequences.[14]

Large reductions in global emissions of heat-trapping gases, similar to the lower emissions scenario (B1) analyzed in this assessment, would reduce the risks of some of the damaging impacts of climate change. Some targets called for in international climate negotiations to date would require even larger reductions than those outlined in the B1 scenario. Meanwhile, global emissions are still rising and are on a path to be even higher than the high emissions scenario (A2) analyzed in this report. The recent U.S. contribution to annual global emissions is about 18%, but the U.S. contribution to cumulative global emissions over the last century is much higher. Carbon dioxide lasts for a long time in the atmosphere, and it is the cumulative carbon emissions that determine the amount of global climate change. After decades of increases, U.S. CO_2 emissions from energy use (which account for 97% of total U.S. emissions) declined by around 9% between 2008 and 2012, largely due to a shift from coal to less CO_2-intensive natural gas for electricity production. Governmental actions in city, state, regional, and federal programs to promote energy efficiency have also contributed to reducing U.S. carbon emissions. Many, if not most of these programs are motivated by other policy objectives, but some are directed specifically at greenhouse gas emissions. These U.S. actions and others that might be undertaken in the future are described in the Mitigation chapter of this report. Over the remainder of this century, aggressive and sustained greenhouse gas emission reductions by the United States and by other nations would be needed to reduce global emissions to a level consistent with the lower scenario (B1) analyzed in this assessment.[15]

With regard to adaptation, the pace and magnitude of observed and projected changes emphasize the need to be prepared for a wide variety and intensity of impacts. Because of the growing influence of human activities, the climate of the past is not a good basis for future planning. For example, building codes and landscaping ordinances could be updated to improve energy efficiency, conserve water supplies, protect against insects that spread disease (such as dengue fever), reduce susceptibility to heat stress, and improve protection against extreme events. The fact that climate change impacts are increasing points to the urgent need to develop and refine approaches that enable decision-making and increase flexibility and resilience in the face of ongoing and future impacts. Reducing non-climate-related stresses that contribute to existing vulnerabilities can also be an effective approach to climate change adaptation.[16]

Adaptation can involve considering local, state, regional, national, and international jurisdictional objectives. For example, in managing water supplies to adapt to a changing climate, the implications of international treaties should be considered in the context of managing the Great Lakes, the Columbia River, and the Colorado River to deal with increased drought risk. Both "bottom up" communi-

ty planning and "top down" national strategies may help regions deal with impacts such as increases in electrical brownouts, heat stress, floods, and wildfires.[17]

Proactively preparing for climate change can reduce impacts while also facilitating a more rapid and efficient response to changes as they happen. Such efforts are beginning at the federal, regional, state, tribal, and local levels, and in the corporate and non-governmental sectors, to build adaptive capacity and resilience to climate change impacts. Using scientific information to prepare for climate changes in advance can provide economic opportunities, and proactively managing the risks can reduce impacts and costs over time.[18]

There are a number of areas where improved scientific information or understanding would enhance the capacity to estimate future climate change impacts. For example, knowledge of the mechanisms controlling the rate of ice loss in Greenland and Antarctica is limited, making it difficult for scientists to narrow the range of expected future sea level rise. Improved understanding of ecological and social responses to climate change is needed, as is understanding of how ecological and social responses will interact.[19]

A sustained climate assessment process could more efficiently collect and synthesize the rapidly evolving science and help supply timely and relevant information to decision-makers. Results from all of these efforts could continue to deepen our understanding of the interactions of human and natural systems in the context of a changing climate, enabling society to effectively respond and prepare for our future.[20]

The cumulative weight of the scientific evidence contained in this report confirms that climate change is affecting the American people now, and that choices we make will affect our future and that of future generations.

Cities providing transportation options including bike lanes, buildings designed with energy saving features such as green roofs, and houses elevated to allow storm surges to pass underneath are among the many response options being pursued around the country.

REPORT FINDINGS

These findings distill important results that arise from this National Climate Assessment. They do not represent a full summary of all of the chapters' findings, but rather a synthesis of particularly noteworthy conclusions.

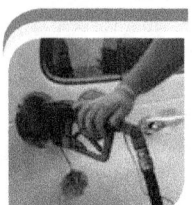

1. Global climate is changing and this is apparent across the United States in a wide range of observations. The global warming of the past 50 years is primarily due to human activities, predominantly the burning of fossil fuels.

Many independent lines of evidence confirm that human activities are affecting climate in unprecedented ways. U.S. average temperature has increased by 1.3°F to 1.9°F since record keeping began in 1895; most of this increase has occurred since about 1970. The most recent decade was the warmest on record. Because human-induced warming is superimposed on a naturally varying climate, rising temperatures are not evenly distributed across the country or over time.[21] See page 18.

2. Some extreme weather and climate events have increased in recent decades, and new and stronger evidence confirms that some of these increases are related to human activities.

Changes in extreme weather events are the primary way that most people experience climate change. Human-induced climate change has already increased the number and strength of some of these extreme events. Over the last 50 years, much of the United States has seen an increase in prolonged periods of excessively high temperatures, more heavy downpours, and in some regions, more severe droughts.[22] See page 24.

3. Human-induced climate change is projected to continue, and it will accelerate significantly if global emissions of heat-trapping gases continue to increase.

Heat-trapping gases already in the atmosphere have committed us to a hotter future with more climate-related impacts over the next few decades. The magnitude of climate change beyond the next few decades depends primarily on the amount of heat-trapping gases that human activities emit globally, now and in the future.[23] See page 28.

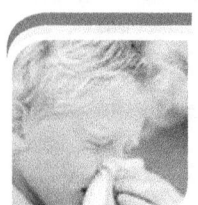

4. Impacts related to climate change are already evident in many sectors and are expected to become increasingly disruptive across the nation throughout this century and beyond.

Climate change is already affecting societies and the natural world. Climate change interacts with other environmental and societal factors in ways that can either moderate or intensify these impacts. The types and magnitudes of impacts vary across the nation and through time. Children, the elderly, the sick, and the poor are especially vulnerable. There is mounting evidence that harm to the nation will increase substantially in the future unless global emissions of heat-trapping gases are greatly reduced.[24] See page 32.

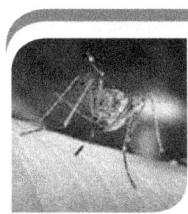

5. Climate change threatens human health and well-being in many ways, including through more extreme weather events and wildfire, decreased air quality, and diseases transmitted by insects, food, and water.

Climate change is increasing the risks of heat stress, respiratory stress from poor air quality, and the spread of waterborne diseases. Extreme weather events often lead to fatalities and a variety of health impacts on vulnerable populations, including impacts on mental health, such as anxiety and post-traumatic stress disorder. Large-scale changes in the environment due to climate change and extreme weather events are increasing the risk of the emergence or reemergence of health threats that are currently uncommon in the United States, such as dengue fever.[25] See page 34.

6. Infrastructure is being damaged by sea level rise, heavy downpours, and extreme heat; damages are projected to increase with continued climate change.

Sea level rise, storm surge, and heavy downpours, in combination with the pattern of continued development in coastal areas, are increasing damage to U.S. infrastructure including roads, buildings, and industrial facilities, and are also increasing risks to ports and coastal military installations. Flooding along rivers, lakes, and in cities following heavy downpours, prolonged rains, and rapid melting of snowpack is exceeding the limits of flood protection infrastructure designed for historical conditions. Extreme heat is damaging transportation infrastructure such as roads, rail lines, and airport runways.[26] See page 38.

7. Water quality and water supply reliability are jeopardized by climate change in a variety of ways that affect ecosystems and livelihoods.

Surface and groundwater supplies in some regions are already stressed by increasing demand for water as well as declining runoff and groundwater recharge. In some regions, particularly the southern part of the country and the Caribbean and Pacific Islands, climate change is increasing the likelihood of water shortages and competition for water among its many uses. Water quality is diminishing in many areas, particularly due to increasing sediment and contaminant concentrations after heavy downpours.[27] See page 42.

8. Climate disruptions to agriculture have been increasing and are projected to become more severe over this century.

Some areas are already experiencing climate-related disruptions, particularly due to extreme weather events. While some U.S. regions and some types of agricultural production will be relatively resilient to climate change over the next 25 years or so, others will increasingly suffer from stresses due to extreme heat, drought, disease, and heavy downpours. From mid-century on, climate change is projected to have more negative impacts on crops and livestock across the country – a trend that could diminish the security of our food supply.[28] See page 46.

9. Climate change poses particular threats to Indigenous Peoples' health, well-being, and ways of life.

Chronic stresses such as extreme poverty are being exacerbated by climate change impacts such as reduced access to traditional foods, decreased water quality, and increasing exposure to health and safety hazards. In parts of Alaska, Louisiana, the Pacific Islands, and other coastal locations, climate change impacts (through erosion and inundation) are so severe that some communities are already relocating from historical homelands to which their traditions and cultural identities are tied. Particularly in Alaska, the rapid pace of temperature rise, ice and snow melt, and permafrost thaw are significantly affecting critical infrastructure and traditional livelihoods.[29] See page 48.

10. Ecosystems and the benefits they provide to society are being affected by climate change. The capacity of ecosystems to buffer the impacts of extreme events like fires, floods, and severe storms is being overwhelmed.

Climate change impacts on biodiversity are already being observed in alteration of the timing of critical biological events such as spring bud burst and substantial range shifts of many species. In the longer term, there is an increased risk of species extinction. These changes have social, cultural, and economic effects. Events such as droughts, floods, wildfires, and pest outbreaks associated with climate change (for example, bark beetles in the West) are already disrupting ecosystems. These changes limit the capacity of ecosystems, such as forests, barrier beaches, and wetlands, to continue to play important roles in reducing the impacts of these extreme events on infrastructure, human communities, and other valued resources.[30] See page 50.

11. Ocean waters are becoming warmer and more acidic, broadly affecting ocean circulation, chemistry, ecosystems, and marine life.

More acidic waters inhibit the formation of shells, skeletons, and coral reefs. Warmer waters harm coral reefs and alter the distribution, abundance, and productivity of many marine species. The rising temperature and changing chemistry of ocean water combine with other stresses, such as overfishing and coastal and marine pollution, to alter marine-based food production and harm fishing communities.[31] See page 58.

12. Planning for adaptation (to address and prepare for impacts) and mitigation (to reduce future climate change, for example by cutting emissions) is becoming more widespread, but current implementation efforts are insufficient to avoid increasingly negative social, environmental, and economic consequences.

Actions to reduce emissions, increase carbon uptake, adapt to a changing climate, and increase resilience to impacts that are unavoidable can improve public health, economic development, ecosystem protection, and quality of life.[32] See page 62.

SUPPORTING EVIDENCE FOR THE REPORT FINDINGS

Icons at the lower left corner of each report finding indicate the chapters drawn on for that section.

CLIMATE TRENDS

These two pages present the Key Messages from the "Our Changing Climate" chapter of the full report. They pertain to Report Findings 1, 2, and 3, evidence for which appears on the following pages.

Global climate is changing and this change is apparent across a wide range of observations. The global warming of the past 50 years is primarily due to human activities. Global climate is projected to continue to change over this century and beyond. The magnitude of climate change beyond the next few decades depends primarily on the amount of heat-trapping gases emitted globally, and how sensitive the Earth's climate is to those emissions.

Temperature

U.S. average temperature has increased by 1.3°F to 1.9°F since record keeping began in 1895; most of this increase has occurred since about 1970. The most recent decade was the nation's warmest on record. Temperatures in the United States are expected to continue to rise. Because human-induced warming is superimposed on a naturally varying climate, the temperature rise has not been, and will not be, uniform or smooth across the country or over time.

Extreme Weather

There have been changes in some types of extreme weather events over the last several decades. Heat waves have become more frequent and intense, especially in the West. Cold waves have become less frequent and intense across the nation. There have been regional trends in floods and droughts. Droughts in the Southwest and heat waves everywhere are projected to become more intense, and cold waves less intense everywhere.

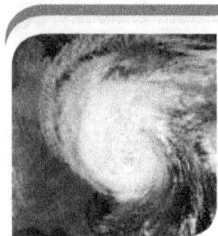

Hurricanes

The intensity, frequency, and duration of North Atlantic hurricanes, as well as the frequency of the strongest (Category 4 and 5) hurricanes, have all increased since the early 1980s. The relative contributions of human and natural causes to these increases are still uncertain. Hurricane-associated storm intensity and rainfall rates are projected to increase as the climate continues to warm.

Severe Storms

Winter storms have increased in frequency and intensity since the 1950s, and their tracks have shifted northward over the United States. Other trends in severe storms, including the intensity and frequency of tornadoes, hail, and damaging thunderstorm winds, are uncertain and are being studied intensively.

Precipitation

Average U.S. precipitation has increased since 1900, but some areas have had increases greater than the national average, and some areas have had decreases. More winter and spring precipitation is projected for the northern United States, and less for the Southwest, over this century.

Heavy Downpours

Heavy downpours are increasing nationally, especially over the last three to five decades. Largest increases are in the Midwest and Northeast. Increases in the frequency and intensity of extreme precipitation events are projected for all U.S. regions.

Frost-free Season

The length of the frost-free season (and the corresponding growing season) has been increasing nationally since the 1980s, with the largest increases occurring in the western United States, affecting ecosystems and agriculture. Across the United States, the growing season is projected to continue to lengthen.

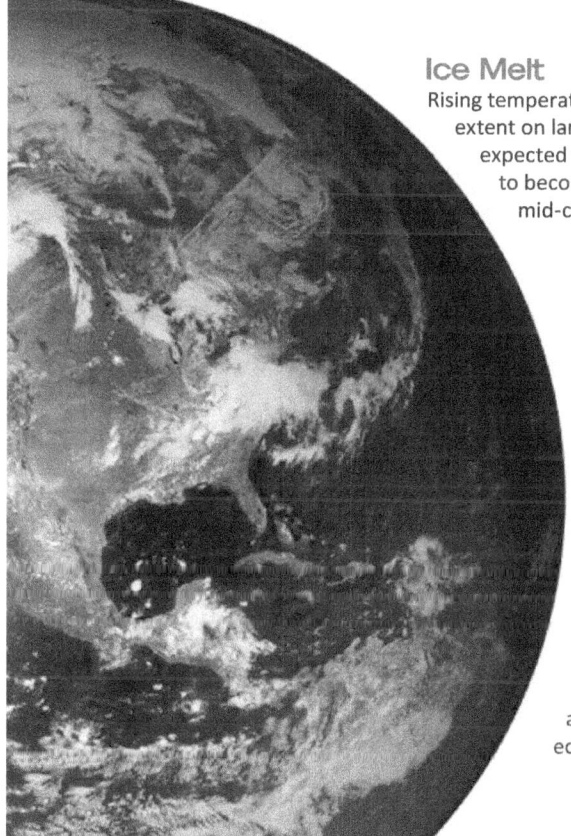

Ice Melt

Rising temperatures are reducing ice volume and surface extent on land, lakes, and sea. This loss of ice is expected to continue. The Arctic Ocean is expected to become essentially ice free in summer before mid-century.

Sea Level

Global sea level has risen by about 8 inches since reliable record keeping began in 1880. It is projected to rise another 1 to 4 feet by 2100.

Ocean Acidification

The oceans are currently absorbing about a quarter of the carbon dioxide emitted to the atmosphere annually and are becoming more acidic as a result, leading to concerns about intensifying impacts on marine ecosystems. See page 60.

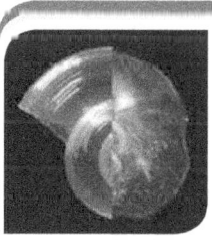

1

OUR CHANGING CLIMATE

Global climate is changing and this is apparent across a wide range of observations.

Evidence for changes in Earth's climate can be found from the top of the atmosphere to the depths of the oceans. Researchers from around the world have compiled this evidence using satellites, weather balloons, thermometers at surface stations, and many other types of observing systems that monitor the Earth's weather and climate. The sum total of this evidence tells an unambiguous story: the planet is warming.

Temperatures at Earth's surface, in the troposphere (the active weather layer extending up to about 5 to 10 miles above the ground), and in the oceans have all increased over recent decades. The largest increases in temperature are occurring closer to the poles, especially in the Arctic. This warming has triggered many other changes to the Earth's climate. Snow and ice cover have decreased in most areas. Atmospheric water vapor is increasing in the lower atmosphere because a warmer atmosphere can hold more water. Sea level is increasing because water expands as it warms and because melting ice on land adds water to the oceans. Changes in other climate-relevant indicators such as growing season length have been observed in many areas. Worldwide, the observed changes in average conditions have been accompanied by increasing trends in extremes of heat and heavy precipitation events, and decreases in extreme cold. It is the sum total of these indicators that leads to the conclusion that warming of our planet is unequivocal.

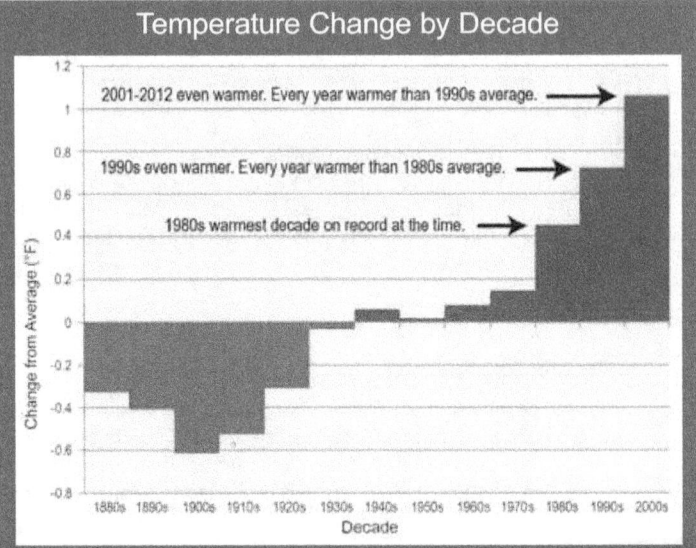

Temperature Change by Decade

2001-2012 even warmer. Every year warmer than 1990s average.

1990s even warmer. Every year warmer than 1980s average.

1980s warmest decade on record at the time.

Change from Average (°F) — vertical axis: 1.2, 1, 0.8, 0.6, 0.4, 0.2, 0, -0.2, -0.4, -0.6, -0.8

Decade — horizontal axis: 1880s 1890s 1900s 1910s 1920s 1930s 1940s 1950s 1960s 1970s 1980s 1990s 2000s

The last five decades have seen a progressive rise in the Earth's average surface temperature. Bars show the difference between each decade's average temperature and the overall average for 1901-2000. (Figure source: NOAA NCDC).

Global Temperature and Carbon Dioxide

Global annual average temperature (as measured over both land and oceans) has increased by more than 1.5°F (0.8°C) since 1880 (through 2012). Red bars show temperatures above the long-term average, and blue bars indicate temperatures below the long-term average. The black line shows atmospheric carbon dioxide (CO_2) concentration in parts per million (ppm). While there is a clear long-term global warming trend, some years do not show a temperature increase relative to the previous year, and some years show greater changes than others. These year-to-year fluctuations in temperature are due to natural processes, such as the effects of El Niños, La Niñas, and volcanic eruptions. (Figure source: updated from Karl et al. 2009[1]).

Global Temperature (°F): 56.0, 56.5, 57.0, 57.5, 58.0
CO_2 Concentration (ppm): 260, 280, 300, 320, 340, 360, 380, 400
Year: 1880, 1900, 1920, 1940, 1960, 1980, 2000

CO_2 Concentration

FAQ

Sea ice in the Arctic has decreased dramatically since the satellite record began in 1978. Minimum Arctic sea ice extent (which occurs in early to mid-September) has decreased by more than 40%.[2] This decline is unprecedented in the historical record, and the reduction of ice volume and thickness is even greater. Ice thickness decreased by more than 50% from 1958-1976 to 2003-2008.[3] The percentage of the March ice cover made up of thicker ice (ice that has survived a summer melt season) decreased from 75% in the mid-1980s to 45% in 2011.[4]

Arctic Sea Ice Decline

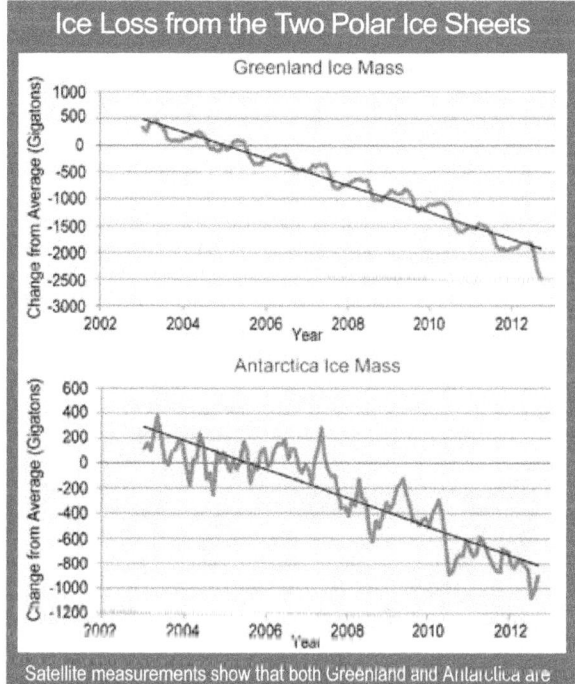

The retreat of sea ice has occurred faster than climate models had predicted. Image on left shows Arctic minimum sea ice extent in 1984, which was about 2.59 million square miles, the average minimum extent for 1979-2000. Image on right shows that the extent of sea ice had dropped to 1.32 million square miles at the end of summer 2012. The dramatic loss of Arctic sea ice increases warming and has many other impacts on the region. Marine mammals including polar bears and many seal species depend on sea ice for nearly all aspects of their existence. Alaska Native coastal communities rely on sea ice for many reasons, including its role as a buffer against coastal erosion from storms and as a platform for hunting. (Figure source: NASA Earth Observatory 2012[8]).

Ice loss increases Arctic warming by replacing white, reflective ice with dark water that absorbs more energy from the sun. More open water can also increase snowfall over northern land areas[5] and increase the north-south meanders of the jet stream, consistent with the occurrence of unusually cold and snowy winters at mid-latitudes in several recent years.[5,6] Significant uncertainties remain in interpreting the effect of Arctic ice changes on mid-latitude weather patterns.[7]

In addition to the rapid decline of Arctic sea ice, rising temperatures are reducing the volume and surface extent of ice on land and lakes. Snow cover on land has also decreased over the past several decades, especially in late spring.

The ice sheets on Greenland and Antarctica are losing mass, adding to global sea level rise.

Ice Loss from the Two Polar Ice Sheets

Satellite measurements show that both Greenland and Antarctica are losing ice as the atmosphere and oceans warm. Melting of the polar ice sheets and glaciers on land add water to the oceans and raise sea level. How fast these two polar ice sheets melt will largely determine how quickly sea level rises. (Figure source: adapted from Wouters et al. 2013[9]).

Climate in the United States is changing.

U.S. average temperature has increased by 1.3°F to 1.9°F since record keeping began in 1895; most of this increase has occurred since about 1970. The most recent decade was the nation's warmest on record. Because human-induced warming is super-imposed on a naturally varying climate, the temperature rise has not been, and will not be, uniform or smooth across the country or over time.

While surface air temperature is the most widely cited measure of climate change, other aspects of climate that are affected by temperature are often more directly relevant to both human society and the natural environment. Examples include shorter duration of ice on lakes and rivers, reduced glacier extent, earlier melting of snowpack, reduced lake levels due to increased evaporation, lengthening of the growing season, changes in plant hardiness zones, increased humidity, rising ocean temperatures, rising sea level, and changes in some types of extreme weather.

Taken as a whole, these changes provide compelling evidence that increasing temperatures are affecting both ecosystems and human society.

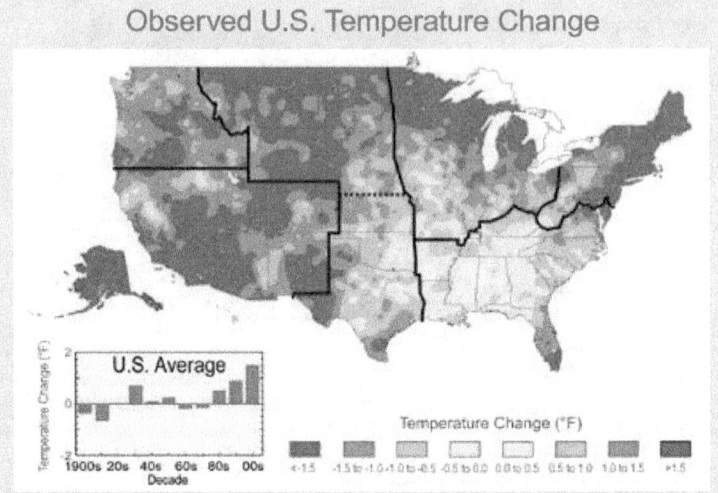

Observed U.S. Temperature Change

The colors on the map show temperature changes over the past 22 years (1991-2012) compared to the 1901-1960 average for the contiguous U.S., and to the 1951-1980 average for Alaska and Hawai'i. The bars on the graph show the average temperature changes for the U.S. by decade for 1901-2012 (relative to the 1901-1960 average). The far right bar (2000s decade) includes 2011 and 2012. The period from 2001 to 2012 was warmer than any previous decade in every region. (Figure source: NOAA NCDC / CICS-NC).

A longer growing season provides a longer period for plant growth and productivity and can slow the increase in atmospheric CO_2 concentrations through increased CO_2 uptake by living things and their environment.[10] The longer growing season can increase the growth of beneficial plants (such as crops and forests) as well as undesirable ones (such as ragweed).[11] In some cases where moisture is limited, the greater evaporation and loss of moisture through plant transpiration (release of water from plant leaves) associated with a longer growing season can mean less productivity because of increased drying[12] and earlier and longer fire seasons.

On the left is a photograph of Muir Glacier in Alaska taken on August 13, 1941; on the right, a photograph taken from the same vantage point on August 31, 2004. Total glacial mass has declined sharply around the globe, adding to sea level rise. (Left photo by glaciologist William O. Field; right photo by geologist Bruce F. Molnia of the United States Geological Survey.)

Increased frost-free season length, especially in already hot and moisture-stressed regions like the Southwest, can lead to further heat stress on plants and increased water demands for crops. Higher temperatures and fewer frost-free days during winter can lead to early bud burst or bloom of some perennial plants, resulting in frost damage when cold conditions occur in late spring. In addition, with higher winter temperatures, some agricultural pests can persist year-round, and new pests and diseases may become established.[13]

The lengthening of the frost-free season has been somewhat greater in the western U.S. than the eastern U.S.,[1] increasing by 2 to 3 weeks in the Northwest and Southwest, 1 to 2 weeks in the Midwest, Great Plains, and Northeast, and slightly less than 1 week in the Southeast. These differences mirror the overall trend of more warming in the north and west and less warming in the Southeast.

Average annual precipitation over the U.S. has increased in recent decades, although there are important regional differences. For example, precipitation since 1991 (relative to 1901-1960) increased the most in the Northeast (8%), Midwest (9%), and southern Great Plains (8%), while much of the Southeast and Southwest had a mix of areas of increases and decreases.

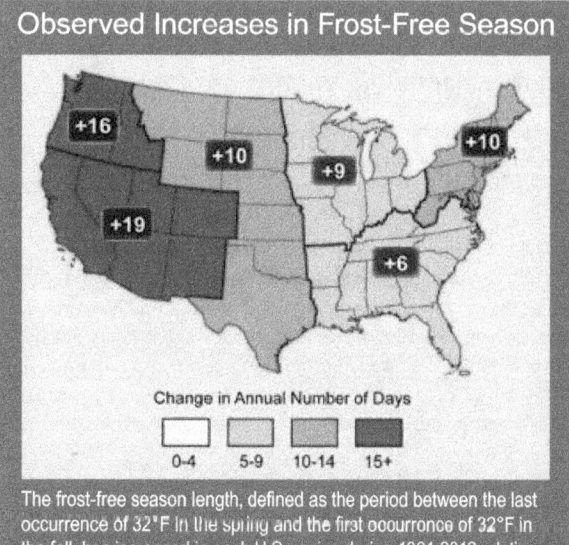

Observed Increases in Frost-Free Season

Change in Annual Number of Days

0-4   5-9   10-14   15+

The frost-free season length, defined as the period between the last occurrence of 32°F in the spring and the first occurrence of 32°F in the fall, has increased in each U.S. region during 1991-2012 relative to 1901-1960. Increases in frost-free season length correspond to similar increases in growing season length. (Figure source: NOAA NCDC / CICS-NC).

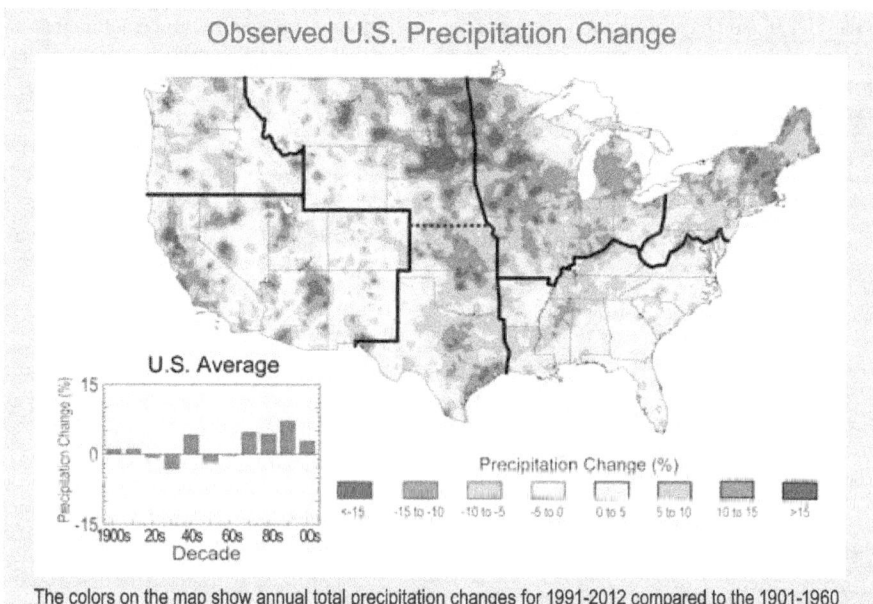

Observed U.S. Precipitation Change

U.S. Average

Precipitation Change (%)

<-15   -15 to -10   -10 to -5   -5 to 0   0 to 5   5 to 10   10 to 15   >15

The colors on the map show annual total precipitation changes for 1991-2012 compared to the 1901-1960 average, and show wetter conditions in most areas. The bars on the graph show average precipitation differences by decade for 1901-2012 (relative to the 1901-1960 average). The far right bar is for 2001-2012. (Figure source: NOAA NCDC / CICS-NC).

The global warming of the past 50 years is primarily due to human activities, predominantly the burning of fossil fuels.

Climate has changed naturally throughout Earth's history. However, natural factors cannot explain the recent observed warming.

In the past, climate change was driven exclusively by natural factors: explosive volcanic eruptions that injected reflective particles into the upper atmosphere, changes in energy from the sun, periodic variations in the Earth's orbit, natural cycles that transfer heat between the ocean and the atmosphere, and slowly changing natural variations in heat-trapping gases in the atmosphere.

All of these natural factors, and their interactions with each other, have altered global average temperature over periods ranging from months to thousands of years. For example, past glacial periods were initiated by shifts in the Earth's orbit, and then amplified by resulting decreases in atmospheric levels of carbon dioxide and subsequently by greater reflection of the sun's energy by ice and snow as the Earth's climate system responded to a cooler climate.

Natural factors are still affecting the planet's climate today. The difference is that, since the beginning of the Industrial Revolution, humans have been increasingly affecting global climate, to the point where we are now the primary cause of recent and projected future change.

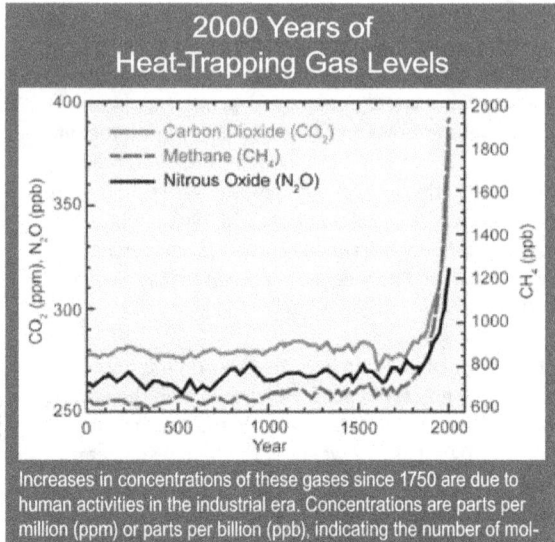

2000 Years of Heat-Trapping Gas Levels

Increases in concentrations of these gases since 1750 are due to human activities in the industrial era. Concentrations are parts per million (ppm) or parts per billion (ppb), indicating the number of molecules of the greenhouse gas per million or billion molecules of air. (Figure source: Forster et al. 2007[14]).

The majority of the warming at the global scale over the past 50 years can only be explained by the effects of human influences, especially the emissions from burning fossil fuels (coal, oil, and natural gas) and from deforestation.

The emissions from human influences affecting climate include heat-trapping gases such as carbon dioxide (CO_2), methane, and nitrous oxide, and particles such as black carbon (soot), which has a warming influence, and sulfates, which have an overall cooling influence. In addition to human-induced global climate change, local climate can also be affected by other human factors (such as crop irrigation) and natural variability.

Carbon dioxide has been building up in the atmosphere since the beginning of the industrial era in the mid-1700s, primarily due to burning coal, oil, and gas, and secondarily due to clearing of forests. Atmospheric levels have increased by about 40% relative to pre-industrial levels.

Methane levels in the atmosphere have increased due to human activities including agriculture (with livestock producing methane in their digestive tracts and rice farming producing it via bacteria that live in the flooded fields); mining coal, extraction and transport of natural gas, and other fossil fuel-related activities;

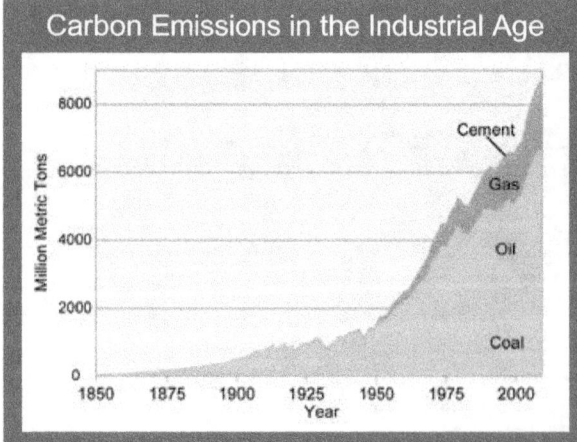

Carbon Emissions in the Industrial Age

Carbon emissions from burning coal, oil, and gas and producing cement, in units of million metric tons of carbon. These emissions account for about 80% of the total emissions of carbon from human activities, with land-use changes (like cutting down forests) accounting for the other 20% in recent decades. (Data from Boden et al. 2012[15]).

and waste disposal including sewage and decomposing garbage in landfills. Since pre-industrial times, methane levels have increased by 250%.

Other heat-trapping gases produced by human activities include nitrous oxide, halocarbons, and ozone. Nitrous oxide levels are increasing, primarily as a result of fertilizer use and fossil fuel burning. The concentration of nitrous oxide has increased by about 20% relative to pre-industrial times.

The conclusion that human influences are the primary driver of recent climate change is based on multiple lines of independent evidence. The first line of evidence is our fundamental understanding of how certain gases trap heat, how the climate system responds to increases in these gases, and how other human and natural factors influence climate. The second line of evidence is from reconstructions of past climates using evidence such as tree rings, ice cores, and corals. These show that global surface temperatures over the last several decades are clearly unusual, with the last decade (2000-2009) warmer than any time in at least the last 1,300 years and perhaps much longer.

The third line of evidence comes from using climate models to simulate the climate of the past century, separating the human and natural factors that influence climate. When the human factors are removed, these models show that solar and volcanic activity would have tended to slightly cool the earth, and other natural variations are too small to explain the amount of warming. Only when the human influences are included do the models reproduce the warming observed over the past 50 years.

Another line of evidence involves so-called "fingerprint" studies that are able to attribute observed climate changes to particular causes. For example, the fact that the stratosphere (the layer above the troposphere) is cooling while the Earth's surface and lower atmosphere are warming is a fingerprint that the warming is due to increases in heat-trapping gases. In contrast, if the observed warming had been due to increases in solar output, Earth's atmosphere would have warmed throughout its entire extent, including the stratosphere. In addition to such temperature analyses, scientific attribution of observed changes to human influence extends to many other aspects of climate, such as changing patterns in precipitation, increasing humidity, changes in pressure, and increasing ocean heat content.

Oil used for transportation and coal used for electricity generation are the largest contributors to the rise in carbon dioxide that is the primary driver of recent climate change.

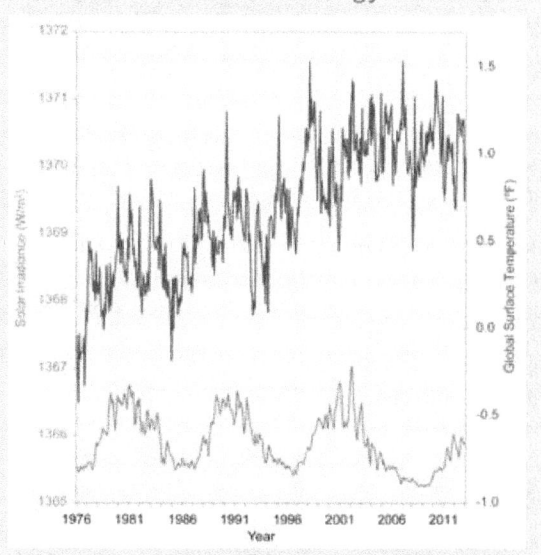

The full record of satellite measurements of the sun's energy received at the top of the Earth's atmosphere is shown in red, following its natural 11-year cycle of small ups and downs, without any net increase. Over the same period, global temperature relative to 1961-1990 average (shown in blue) has risen markedly. This is a clear indication that changes in the sun are not responsible for the observed warming over recent decades. (Figure source: NOAA NCDC / CICS-NC).

FINDING

2 EXTREME WEATHER

Some extreme weather and climate events have increased in recent decades, and new and stronger evidence confirms that some of these increases are related to human activities.

As the world has warmed, that warming has triggered many other changes to the Earth's climate. Changes in extreme weather and climate events, such as heat waves and droughts, are the primary way that most people experience climate change. Human-induced climate change has already increased the number and strength of some of these extreme events. Over the last 50 years, much of the U.S. has seen increases in prolonged periods of excessively high temperatures, heavy downpours, and in some regions, severe floods and droughts.

Heat Waves

Heat waves are periods of abnormally hot weather lasting days to weeks. The number of heat waves has been increasing in recent years. This trend has continued in 2011 and 2012, with the number of intense heat waves being almost triple the long-term average. The recent heat waves and droughts in Texas (2011) and the Midwest (2012) set records for highest monthly average temperatures. Analyses show that human-induced climate change has generally increased the probability of heat waves.[1] And prolonged (multi-month) extreme heat has been unprecedented since the start of reliable instrumental records in 1895.

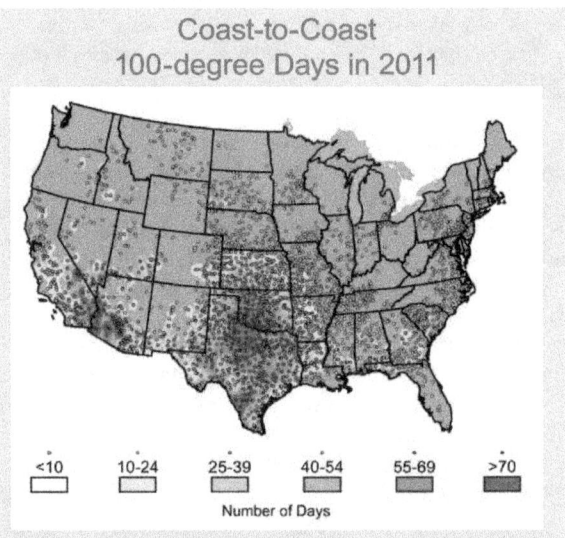

Coast-to-Coast 100-degree Days in 2011

Number of Days: <10, 10-24, 25-39, 40-54, 55-69, >70

Map shows numbers of days with temperatures above 100°F during 2011. Black circles denote the location of observing stations recording at least one such day. The number of days with temperatures exceeding 100°F is expected to increase. The record temperatures and drought during the summer of 2011 represent conditions that will occur more frequently in the U.S. as climate change continues. (Figure source: NOAA NCDC).

Drought

Higher temperatures lead to increased rates of evaporation, including more loss of moisture through plant leaves. Even in areas where precipitation does not decrease, these increases in surface evaporation and loss of water from plants lead to more rapid drying of soils if the effects of higher temperatures are not

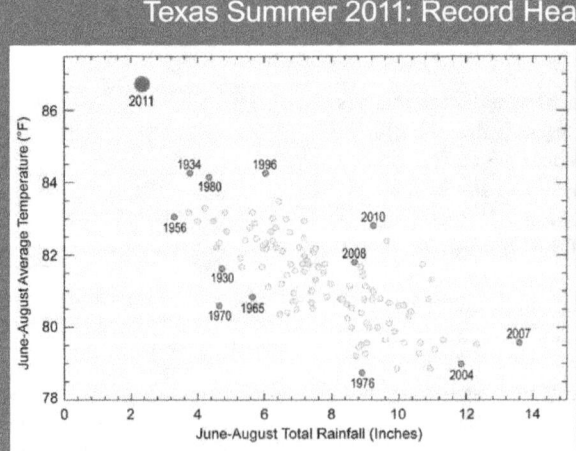

Texas Summer 2011: Record Heat and Drought

Dots show the average summer temperature and total rainfall in Texas for each year from 1895 to 2012. Red dots illustrate the range of temperatures and rainfall observed over time. The record temperatures and drought during the summer of 2011 (large red dot) represent conditions far outside those that have occurred since the instrumental record began.[2] An analysis has shown that the probability of such an event has more than doubled as a result of human-induced climate change.[3] (Figure source: NOAA NCDC / CICS-NC).

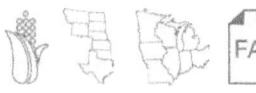

FAQ

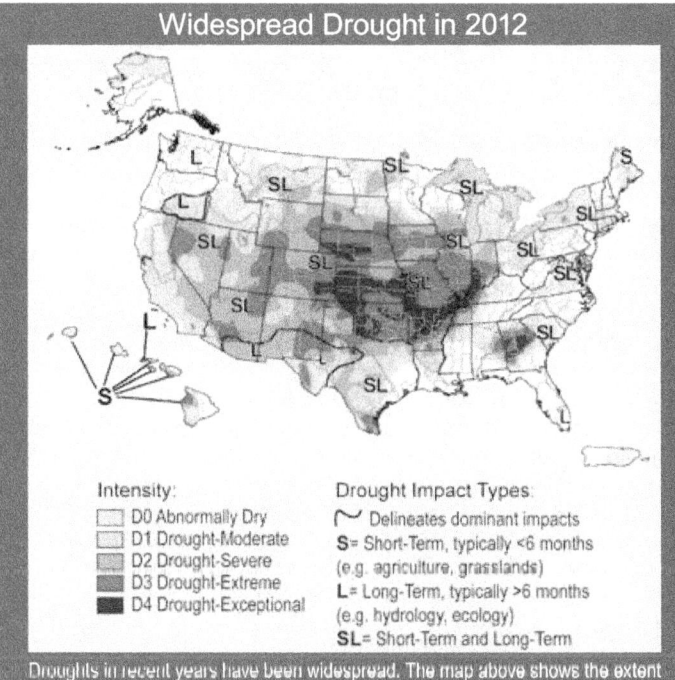

Widespread Drought in 2012

Intensity:
- ☐ D0 Abnormally Dry
- ☐ D1 Drought-Moderate
- ☐ D2 Drought-Severe
- ☐ D3 Drought-Extreme
- ■ D4 Drought-Exceptional

Drought Impact Types:
- ∼ Delineates dominant impacts
- **S**= Short-Term, typically <6 months (e.g. agriculture, grasslands)
- **L**= Long-Term, typically >6 months (e.g. hydrology, ecology)
- **SL**= Short-Term and Long-Term

Droughts in recent years have been widespread. The map above shows the extent of drought in mid August 2012. The U.S. Drought Monitor is produced in partnership between the National Drought Mitigation Center at the University of Nebraska-Lincoln, the United States Department of Agriculture, and the National Oceanic and Atmospheric Administration. (Map courtesy of NDMC-UNL).

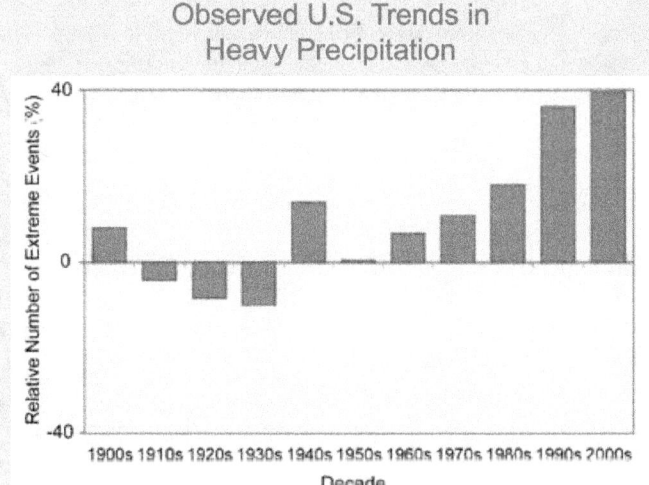

Observed U.S. Trends in Heavy Precipitation

One measure of heavy precipitation events is a two-day precipitation total that is exceeded on average only once in a 5-year period, also known as the once-in-five-year event. As this extreme precipitation index for 1901-2012 shows, the occurrence of such events has become much more common in recent decades. Changes are compared to the period 1901-1960, and do not include Alaska or Hawai'i. (Figure source: adapted from Kunkel et al. 2013[7]).

offset by other changes (such as reduced wind speed or increased humidity).[4] As soil dries out, a larger proportion of the incoming heat from the sun goes into heating the soil and adjacent air rather than evaporating its moisture, resulting in hotter summers under drier climatic conditions.[5]

An example of recent drought occurred in 2011, when many locations in Texas and Oklahoma experienced more than 100 days over 100°F. Both states set new records for the hottest summer since record keeping began in 1895. Rates of water loss, due in part to evaporation, were double the long-term average. The heat and drought depleted water resources and contributed to more than $10 billion in direct losses to agriculture alone.

Heavy Downpours

Heavy downpours are increasing nationally, especially over the last three to five decades. The heaviest rainfall events have become heavier and more frequent, and the amount of rain falling on the heaviest rain days has also increased. Since 1991, the amount of rain falling in very heavy precipitation events has been significantly above average. This increase has been greatest in the Northeast, Midwest, and upper Great Plains – more than 30% above the 1901-1960 average. There has also been an increase in flooding events in the Midwest and Northeast, where the largest increases in heavy rain amounts have occurred.

The mechanism driving these changes is well understood. Warmer air can contain more water vapor than cooler air. Global analyses show that the amount of water vapor in the atmosphere has in fact increased due to human-caused warming.[6] This extra moisture is available to storm systems, resulting in heavier rainfalls. Climate change also alters characteristics of the atmosphere that affect weather patterns and storms.

Floods

Flooding may intensify in many U.S. regions, even in areas where total precipitation is projected to decline. A flood is defined as any high flow, overflow, or inundation by water that causes or threatens damage.[8] Floods are caused or amplified by both weather- and human-related factors. Major weather factors include heavy or prolonged precipitation, snowmelt, thunderstorms, storm surges from hurricanes, and ice or debris jams. Human factors include structural failures of dams and levees, altered drainage, and land-cover alterations (such as pavement).

Increasingly, humanity is also adding to weather-related factors, as human-induced warming increases heavy downpours, causes more extensive storm surges due to sea level rise, and leads to more rapid spring snowmelt.

MAJOR FLOOD TYPES

All flood types are affected by climate-related factors, some more than others.

Flash floods occur in small and steep watersheds and waterways and can be caused by short-duration intense precipitation, dam or levee failure, or collapse of debris and ice jams. Most flood-related deaths in the U.S. are associated with flash floods.

Urban flooding can be caused by short-duration very heavy precipitation. Urbanization creates large areas of impervious surfaces (such as roads, pavement, parking lots, and buildings) that increased immediate runoff, and heavy downpours can exceed the capacity of storm drains and cause urban flooding.

Flash floods and urban flooding are directly linked to heavy precipitation and are expected to increase as a result of increases in heavy precipitation events.

River flooding occurs when surface water drained from a watershed into a stream or a river exceeds channel capacity, overflows the banks, and inundates adjacent low lying areas. Riverine flooding depends on precipitation as well as many other factors, such as existing soil moisture conditions and snowmelt.

Coastal flooding is predominantly caused by storm surges that accompany hurricanes and other storms that push large seawater domes toward the shore. Storm surge can cause deaths, widespread infrastructure damage, and severe beach erosion. Storm-related rainfall can also cause inland flooding and is responsible for more than half of the deaths associated with tropical storms.[8] Climate change affects coastal flooding through sea level rise and storm surge, and increases in heavy rainfall during storms.

Worldwide, from 1980 to 2009, floods caused more than 500,000 deaths and affected more than 2.8 billion people.[9] In the United States, floods caused 4,586 deaths from 1959 to 2005[10] while property and crop damage averaged nearly 8 billion dollars per year (in 2011 dollars) over 1981 through 2011.[8] The risks from future floods are significant, given expanded development in coastal areas and floodplains, unabated urbanization, land-use changes, and human-induced climate change.[9]

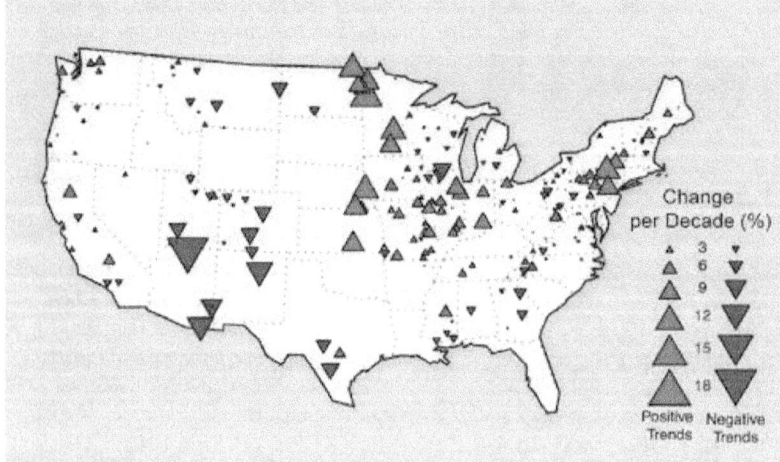

Trends in Flood Magnitude

There are significant trends in the magnitude of river flooding in many parts of the United States.[11] River flood magnitudes (from the 1920s through 2008) have decreased in the Southwest and increased in the eastern Great Plains, parts of the Midwest, and from the northern Appalachians into New England.[12] The map shows increasing trends in floods in green and decreasing trends in brown. The magnitude of these trends is illustrated by the size of the triangles. (Figure source: Peterson et al. 2013[12]).

Change per Decade (%)

3, 6, 9, 12, 15, 18

Positive Trends   Negative Trends

Hurricanes

There has been a substantial increase in most measures of Atlantic hurricane activity since the early 1980s, the period during which high quality satellite data are available.[13] These include measures of intensity, frequency, and duration as well as the number of strongest (Category 4 and 5) storms. The recent increases in activity are linked, in part, to higher sea surface temperatures in the region that Atlantic hurricanes form in and move through. Numerous factors have been shown to influence these local sea surface temperatures, including natural variability, human-induced emissions of heat-trapping gases, and particulate pollution. Quantifying the relative contributions of natural and human-caused factors is an active focus of research.

Hurricane development, however, is influenced by more than just sea surface temperature. How hurricanes develop also depends on how the local atmosphere responds to changes in local sea surface temperatures, and this atmospheric response depends critically on the cause of the change.[14] For example, the atmosphere responds differently when local sea surface temperatures increase due to a local decrease of particulate pollution that allows more sunlight through to warm the ocean, versus when sea surface temperatures increase more uniformly around the world due to increased amounts of human-caused heat-trapping gases.[15]

By late this century, models, on average, project an increase in the number of the strongest (Category 4 and 5) hurricanes. Models also project greater rainfall rates in hurricanes in a warmer climate, with increases of about 20% averaged near the center of hurricanes.

Change in Other Storms

Winter storms have increased in frequency and intensity since the 1950s,[16] and their tracks have shifted northward over the United States.[17] Other trends in severe storms, including the intensity and frequency of tornadoes, hail, and damaging thunderstorm winds, are uncertain and are being studied intensively. There has been a sizable upward trend in the number of storms causing large financial and other losses.[18] However, there are societal contributions to this trend, such as increases in population and wealth.[7]

North Atlantic hurricanes have increased in intensity, frequency, and duration since the early 1980s.

Storm surges reach farther inland as they ride on top of sea levels that are higher due to warming.

Heavy snowfalls during winter storms affect transportation systems and other infrastructure.

FINDING
3 FUTURE CLIMATE

Human-induced climate change is projected to continue, and it will accelerate significantly if emissions of heat-trapping gases continue to increase.

Heat-trapping gases already in the atmosphere have committed us to a hotter future with more climate-related impacts over the next few decades. The magnitude of climate change beyond the next few decades depends primarily on the amount of heat-trapping gases that human activities emit globally, now and in the future.

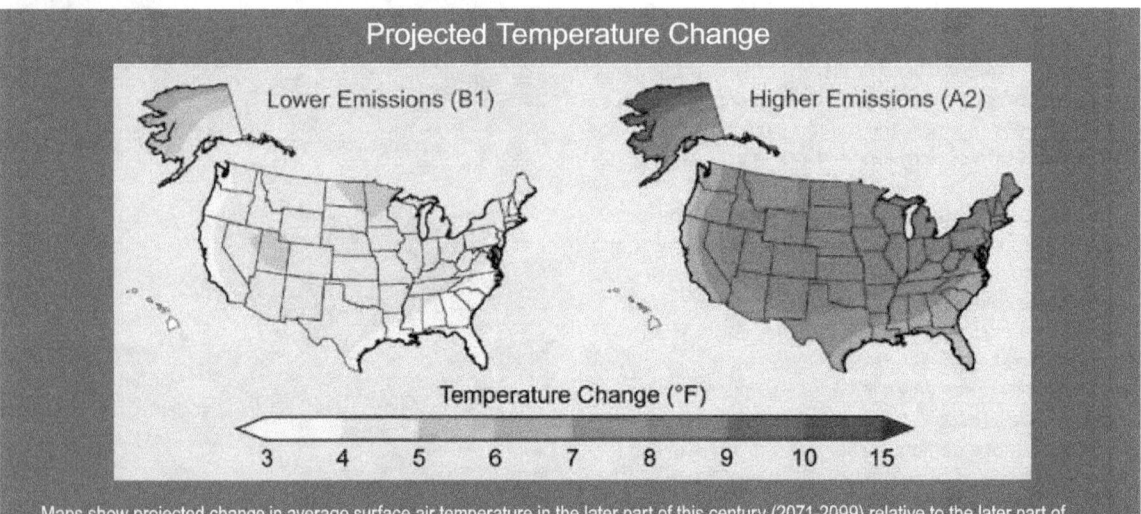

Projected Temperature Change

Lower Emissions (B1)   Higher Emissions (A2)

Temperature Change (°F)

3   4   5   6   7   8   9   10   15

Maps show projected change in average surface air temperature in the later part of this century (2071-2099) relative to the later part of the last century (1970-1999) under a scenario that assumes substantial reductions in heat trapping gases (B1, left) and a higher emissions scenario that assumes continued increases in global emissions (A2, right). These scenarios are used throughout this report for assessing impacts under lower and higher emissions. (Figure source: NOAA NCDC / CICS-NC).

Projected Changes in Soil Moisture

End-of-Century Changes
Lower Emissions Scenario (B1)   Higher Emissions Scenario (A2)

Increased temperatures and changing precipitation patterns will alter soil moisture, which is important for agriculture and ecosystems and has many societal implications. These maps show average change in soil moisture compared to 1971-2000, as projected for late this century (2071-2100) under two emissions scenarios, a lower scenario (B1) and a higher scenario (A2).[1] Eastern U.S. is not displayed because model simulations were only run for the area shown. (Figure source: NOAA NCDC / CICS-NC).

Change (%)

-15 -10 -5 -1   1   5   10   15

U.S. GLOBAL CHANGE RESEARCH PROGRAM

Higher Emissions (A2)

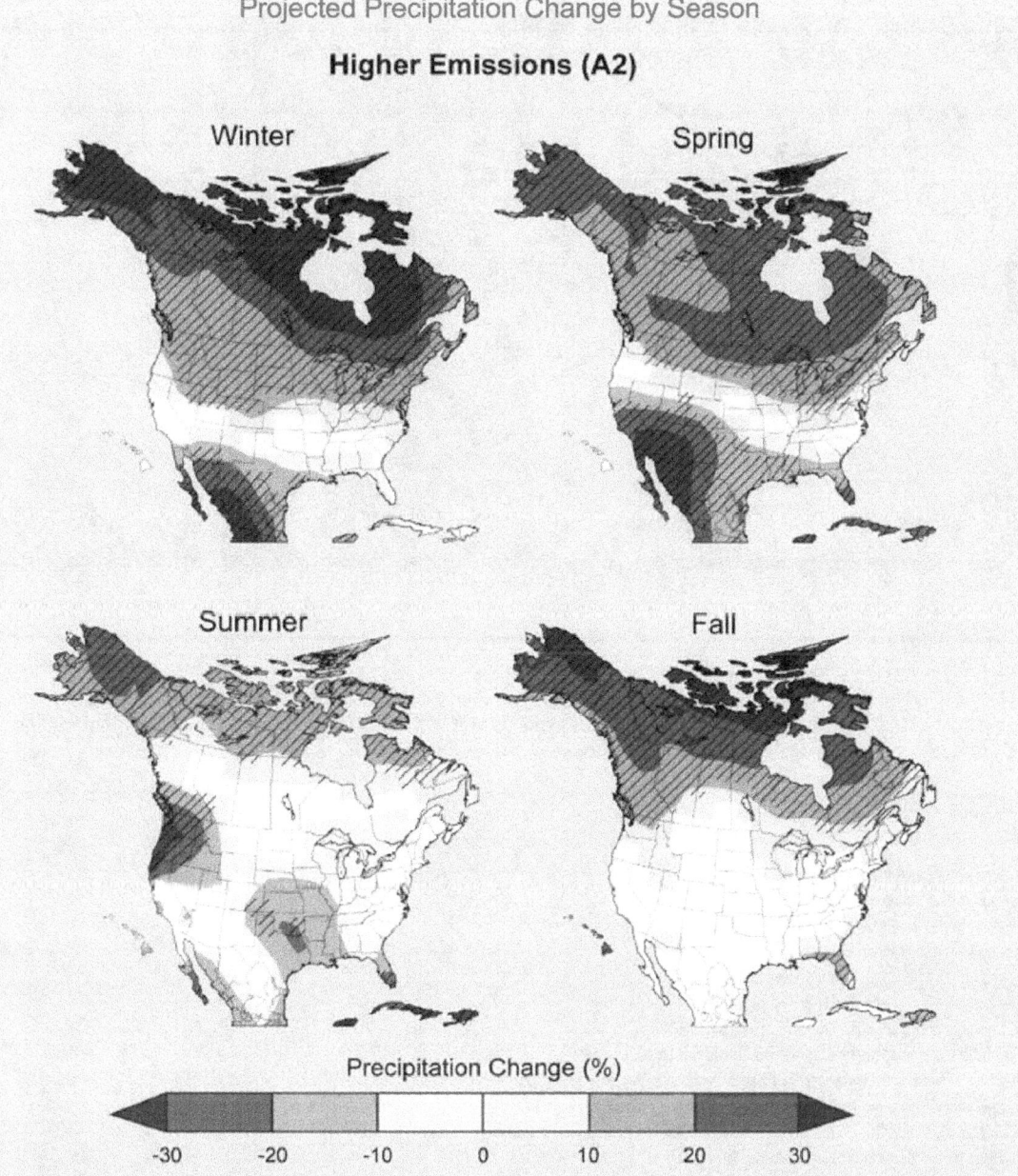

Precipitation Change (%)

-30   -20   -10   0   10   20   30

Climate change affects more than just temperature. The location, timing, and amounts of precipitation will also change as temperatures rise. Maps show projected percent change in precipitation in each season for 2071-2099 (compared to the period 1970-1999) under an emissions scenario that assumes continued increases in emissions (A2). Teal indicates precipitation increases, and brown, decreases. Hatched areas indicate that the projected changes are significant and consistent among models. White areas indicate that the changes are not projected to be larger than could be expected from natural variability. In general, the northern part of the U.S. is projected to see more winter and spring precipitation, while the southwestern U.S. is projected to experience less precipitation in the spring. Wet regions are generally projected to become wetter while dry regions become drier. Summer drying is projected for parts of the U.S., including the Northwest and southern Great Plains. (Figure source: NOAA NCDC / CICS-NC).

Change in Maximum Number of Consecutive Dry Days

Map shows change in the number of consecutive dry days (days receiving less than 0.04 inches of precipitation) at the end of this century (2070-2099) relative to the end of last century (1971-2000) under the highest scenario considered in this report, RCP 8.5. Stippling indicates areas where changes are consistent among at least 80% of the 25 models used in this analysis. (Figure source: NOAA NCDC / CICS-NC).

Change (%)

-20   -15   -10   -5   0   5   10   15   20

Sea level rise

Global sea level has risen about 8 inches since reliable record keeping began in 1880. It is projected to rise another 1 to 4 feet by 2100. The oceans are absorbing over 90% of the increased atmospheric heat associated with emissions from human activity.[2] Like mercury in a thermometer, water expands as it warms up (this is referred to as "thermal expansion") causing sea levels to rise. Melting of glaciers and ice sheets is also contributing to sea level rise at increasing rates.[3]

Past and Projected Changes in Global Sea Level

Figure shows estimated, observed, and possible amounts of global sea level rise from 1800 to 2100, relative to the year 2000. Estimates from proxy data[4] (for example, based on sediment records) are shown in red (1800-1890, pink band shows uncertainty), tide gauge data in blue for 1880-2009,[5] and satellite observations are shown in green from 1993 to 2012.[6] The future scenarios range from 0.66 feet to 6.6 feet in 2100.[7] These scenarios are not based on climate model simulations, but rather reflect the range of possible scenarios based on other kinds of scientific studies. The orange line at right shows the currently projected range of sea level rise of 1 to 4 feet by 2100, which falls within the larger risk-based scenario range. The large projected range reflects uncertainty about how glaciers and ice sheets will react to the warming ocean, the warming atmosphere, and changing winds and currents. As seen in the observations, there are year-to-year variations in the trend. (Figure source: Adapted from Parris et al. 2012,[7] with contributions from NASA Jet Propulsion Laboratory).

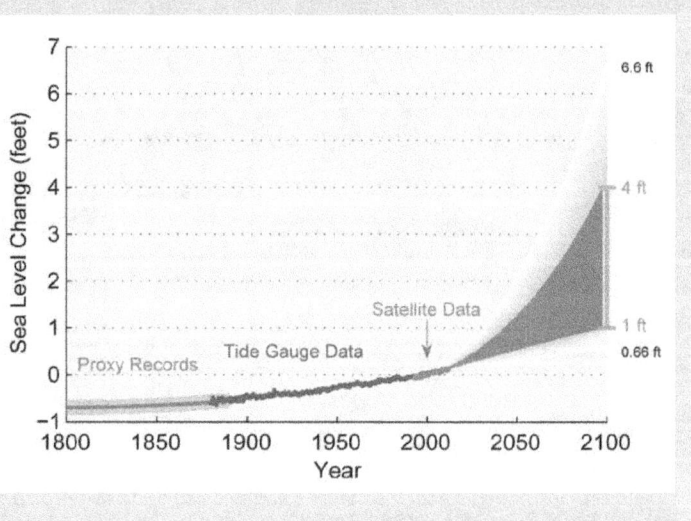

Emission Levels Determine Temperature Rises

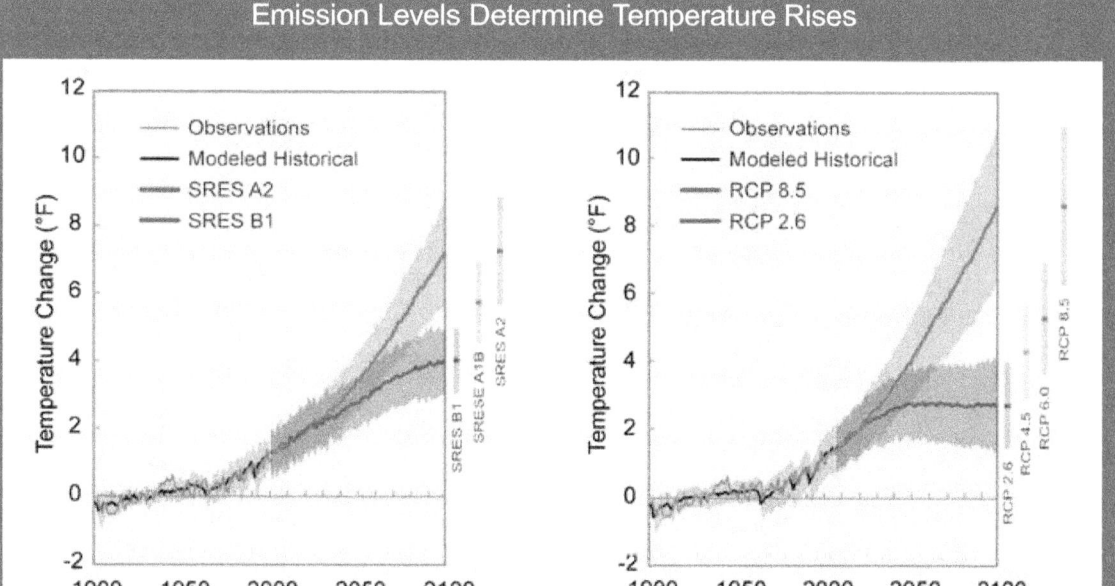

Different amounts of heat-trapping gases released into the atmosphere by human activities produce different projected increases in Earth's temperature. In the figure, each line represents a central estimate of global average temperature rise for a specific emissions pathway (relative to the 1901-1960 average). Shading indicates the range (5th to 95th percentile) of results from a suite of climate models. Projections in 2099 for additional emissions pathways are indicated by the bars to the right of each panel. In all cases, temperatures are expected to rise, although the difference between lower and higher emissions pathways is substantial.

The left panel shows the two main scenarios (SRES) used in this report: A2 assumes continued increases in emissions throughout this century, and B1 assumes significant emissions reductions beginning around 2050, though not due explicitly to climate change policies. The right panel shows newer analyses, which are results from the most recent generation of climate models (CMIP5) using the most recent emissions pathways (RCPs). Some of these new projections explicitly consider climate policies that would result in emissions reductions, which the SRES set did not.[8] The newest set includes both lower and higher pathways than did the previous set. The lowest emissions pathway shown here, RCP 2.6, assumes immediate and rapid reductions in emissions and would result in about 2.5°F of warming in this century. The highest pathway, RCP 8.5, roughly similar to a continuation of the current path of global emissions increases, is projected to lead to more than 8°F warming by 2100, with a high-end possibility of more than 11°F. (Data from CMIP3, CMIP5, and NOAA NCDC).

Where we are heading

Both voluntary activities and a variety of policies and measures that lower emissions are currently in place at federal, state, and local levels in the U.S., even though there is no comprehensive national climate legislation. Over the remainder of this century, aggressive and sustained greenhouse gas emission reductions by the U.S. and by other nations would be needed to reduce global emissions to a level consistent with the lower scenario (B1) analyzed in this assessment.

4 WIDESPREAD IMPACTS

Impacts related to climate change are already evident in many sectors and are expected to become increasingly disruptive across the nation throughout this century and beyond.

Storm surge on top of sea level rise exacerbates coastal flooding during hurricanes.

Climate change is already affecting societies and the natural world. Climate change interacts with other environmental and societal factors in ways that can either moderate or intensify these impacts. The types and magnitudes of impacts vary across the nation and through time. Children, the elderly, the sick, and the poor are especially vulnerable. There is mounting evidence that harm to the nation will increase substantially in the future unless global emissions of heat-trapping gases are greatly reduced.

Because environmental, cultural, and socioeconomic systems are tightly coupled, climate change impacts can either be amplified or reduced by cultural and socioeconomic decisions. In many arenas, it is clear that societal decisions have substantial influence on the vulnerability of valued resources to climate change. For example, rapid population growth and development in coastal areas tends to amplify climate change related impacts. Recognition of these couplings, together with recognition of multiple sources of vulnerability, helps identify what information decision-makers need as they manage risks.

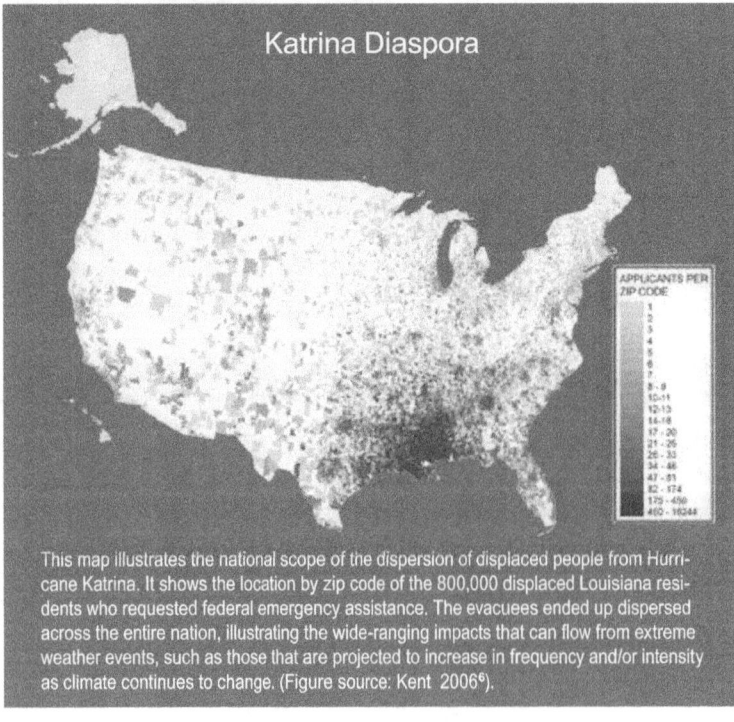

Katrina Diaspora

APPLICANTS PER ZIP CODE

This map illustrates the national scope of the dispersion of displaced people from Hurricane Katrina. It shows the location by zip code of the 800,000 displaced Louisiana residents who requested federal emergency assistance. The evacuees ended up dispersed across the entire nation, illustrating the wide-ranging impacts that can flow from extreme weather events, such as those that are projected to increase in frequency and/or intensity as climate continues to change. (Figure source: Kent 2006[6]).

Multiple System Failures During Extreme Events

Impacts are particularly severe when critical systems simultaneously fail. We have already seen multiple system failures during an extreme weather event in the United States, as when Hurricane Katrina struck New Orleans.[1] Infrastructure and evacuation failures and collapse of critical response services during a storm is one example of multiple system failures. Another example is a loss of electrical power during heat waves or wildfires, which can reduce food and water safety.[2] Air conditioning has helped reduce illness and death due to extreme heat,[3] but if power is lost, everyone is vulnerable. By their nature, such events can exceed our capacity to respond.[4] In succession, these events severely deplete resources needed to respond, from the individual to the national scale, but disproportionately affect the most vulnerable populations.[5]

Coral Reef Ecosystem Collapse

In many social and natural systems, climate change combines with other stresses to cause or expand impacts. For example, coral reefs are threatened by a combination of ocean acidification caused by increased carbon dioxide, rising ocean temperatures, and a variety of other factors caused by human activities.

Recent research indicates that 75% of the world's coral reefs are threatened due to the interactive effects of climate change and local sources of stress, such as overfishing, nutrient pollution, and disease.[7] In Florida, all reefs are rated as threatened; with significant impacts on valuable ecosystem services they provide.[8] Caribbean coral cover has decreased 80% in less than three decades.[9]

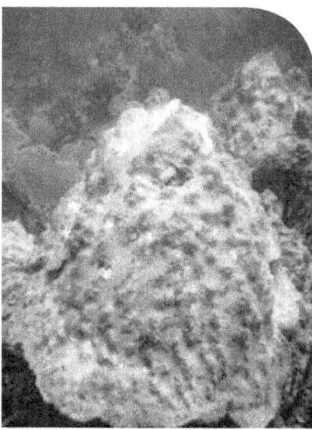

Warm water caused this coral colony to "bleach" (left) as it expelled the symbiotic algae that gave it color and nourishment. The coral then experienced more disease (right), which eventually killed the colony.

These declines have in turn led to a flattening of the three dimensional structure of coral reefs and hence a decrease in the capacity of coral reefs to provide shelter and additional resources for other reef-dependent ocean life.[10]

The relationship between coral and zooxanthellae (algae vital for reef-building corals) is disrupted by higher than usual temperatures and results in a condition where the coral is still alive, but devoid of all its color (bleaching). Bleached corals can later die or become infected with disease.[11] Thus, high temperature events alone can kill large stretches of coral reef, although cold water and poor water quality can also cause localized bleaching and death. Evidence suggests that relatively pristine reefs, with fewer human impacts and with intact fish and associated invertebrate communities, are more resilient to coral bleaching and disease.[12]

Cascading Effects Across Sectors

Agriculture, water, energy, transportation, and more, are all affected by climate change. These sectors of our economy do not exist in isolation and are linked in increasingly complex ways. For example, water supply and energy use are completely intertwined, since water is used to generate energy, and energy is required to pump, treat, and deliver water – which means that irrigation-dependent farmers and urban dwellers are linked as well.

A recent illustration of these interconnections took place during the widespread drought of 2011-2012 when high temperatures caused increased demand for electricity for air conditioning, which resulted in increased water withdrawal and consumption for electricity generation. Heat, increased evaporation, drier soils, and lack of rain led to higher irrigation demands, which added stress on water resources required for energy production. At the same time, low-flowing and warmer rivers threatened to suspend power plant production in several locations, reducing the options for dealing with the concurrent increase in electricity demand.

With electricity demands at all time highs, water shortages threatened more than 3,000 megawatts of generating capacity – enough power to supply more than one million homes.[13] As a result of the record demand and reduced supply, electricity prices spiked.[14]

Heat and drought lead to cascading impacts among sectors including agriculture, water, and energy.

5 HUMAN HEALTH

Climate change threatens human health and well-being in many ways.

Climate change is increasing the risks of respiratory stress from poor air quality, heat stress, and the spread of food-borne, insect-borne, and waterborne diseases. Extreme weather events often lead to fatalities and a variety of health impacts on vulnerable populations, including impacts on mental health, such as anxiety and post-traumatic stress disorder. Large-scale changes in the environment due to climate change and extreme weather events are increasing the risk of the emergence or reemergence of health threats that are currently uncommon in the United States, such as dengue fever.

Key weather and climate drivers of health impacts include increasingly frequent, intense, and longer-lasting extreme heat, which worsens drought, wildfire, and air pollution risks; increasingly frequent extreme precipitation, intense storms, and changes in precipitation patterns that can lead to flooding, drought, and ecosystem changes; and rising sea levels that intensify coastal flooding and storm surge, causing injuries, deaths, stress due to evacuations, and water quality impacts, among other effects on public health.

KEY MESSAGES: HUMAN HEALTH

Climate change threatens human health and well-being in many ways, including impacts from increased extreme weather events, wildfire, decreased air quality, threats to mental health, and illnesses transmitted by food, water, and disease-carriers such as mosquitoes and ticks. Some of these health impacts are already underway in the United States.

Climate change will, absent other changes, amplify some of the existing health threats the nation now faces. Certain people and communities are especially vulnerable, including children, the elderly, the sick, the poor, and some communities of color.

Public health actions, especially preparedness and prevention, can do much to protect people from some of the impacts of climate change. Early action provides the largest health benefits. As threats increase, our ability to adapt to future changes may be limited.

Responding to climate change provides opportunities to improve human health and well-being across many sectors, including energy, agriculture, and transportation. Many of these strategies offer a variety of benefits, protecting people while combating climate change and providing other societal benefits.

Air Quality

Climate change is projected to harm human health by increasing ground-level ozone and/or particulate matter in some locations. Ground-level ozone (a key component of smog) is associated with many health problems, such as diminished lung function, increased hospital admissions and emergency room visits for asthma, and increases in premature deaths.[1] Factors that affect ozone formation include heat, concentrations of precursor chemicals, and methane emissions, while particulate matter concentrations are affected by wildfire emissions and air stagnation episodes, among other factors.[2]

Wildfire Smoke has Widespread Health Effects

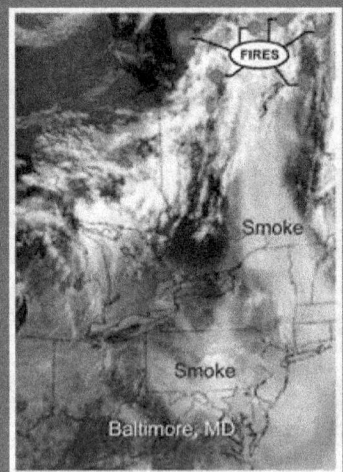

Wildfires, which are projected to increase in some regions due to climate change, have health impacts that can extend hundreds of miles. Forest fires in Quebec, Canada, during July 2002 resulted in up to a 30-fold increase in airborne fine particle concentrations in Baltimore, a city nearly a thousand miles downwind. These fine particles are extremely harmful to human health, affecting both indoor and outdoor air quality. An average of 6.4 million acres burned in U.S. wildfires each year between 2000 and 2010, with 9.5 million acres burned in 2006 and 9.1 million acres in 2012.[3] Global deaths from wildfire smoke have been estimated at 260,000 to 600,000 annually.[4] (Figure source: MODIS instrument on the Terra Satellite, Land Rapid Response Team, NASA/GSFC).

Warmer and drier conditions have already contributed to increasing wildfire extent across the western United States, and future increases are projected in some regions.[5,6] Long periods of record high temperatures are associated with droughts that contribute to dry conditions and drive wildfires in some areas.[7] Wildfire smoke contains particulate matter, carbon monoxide, and other compounds, which can significantly reduce air quality, both locally and in areas downwind of fires.[8,9] Smoke exposure increases respiratory and cardiovascular hospitalizations, emergency room visits and medication for asthma, bronchitis, chest pain, and other ailments.[8,10,11] It has been associated with hundreds of thousands of deaths globally each year.[4,8,10,12] Future climate change is projected to increase wildfire risks and associated emissions, with harmful impacts on health.[6,13]

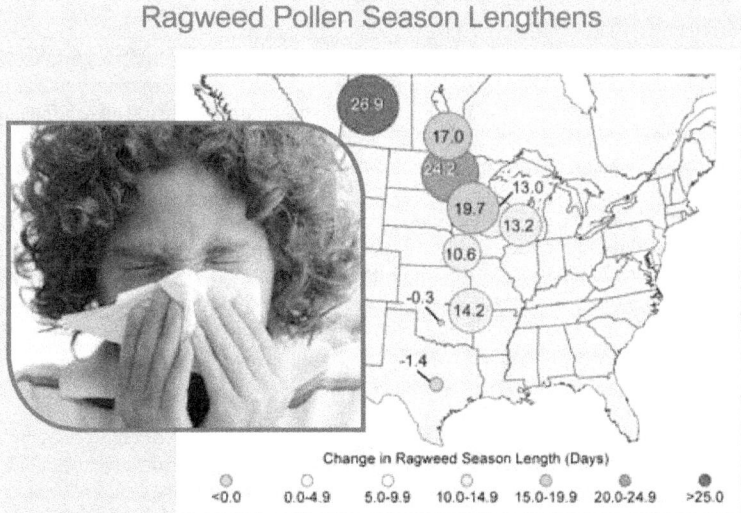

Ragweed Pollen Season Lengthens

Change in Ragweed Season Length (Days)

| <0.0 | 0.0-4.9 | 5.0-9.9 | 10.0-14.9 | 15.0-19.9 | 20.0-24.9 | >25.0 |

Ragweed pollen season length has increased in central North America between 1995 and 2011 by as much as 11 to 27 days in parts of the U.S. and Canada, in response to rising temperatures. Increases in the length of this allergenic pollen season are correlated with increases in the number of days before the first frost. As shown in the figure, the largest increases have been observed in northern cities. (Data updated from Ziska et al. 2011[14]).

Allergies and Asthma

Climate change, as well as increased CO_2 by itself, can contribute to increased production of plant-based allergens.[6,14,15] Higher pollen concentrations and longer pollen seasons can increase allergic sensitizations and asthma episodes,[16,17] and diminish productive work and school days.[14,17,18] Simultaneous exposure to toxic air pollutants can worsen allergic responses.[19] Extreme rainfall and rising temperatures can also foster indoor air quality problems, including the growth of indoor fungi and molds, with increases in respiratory and asthma-related conditions.[20]

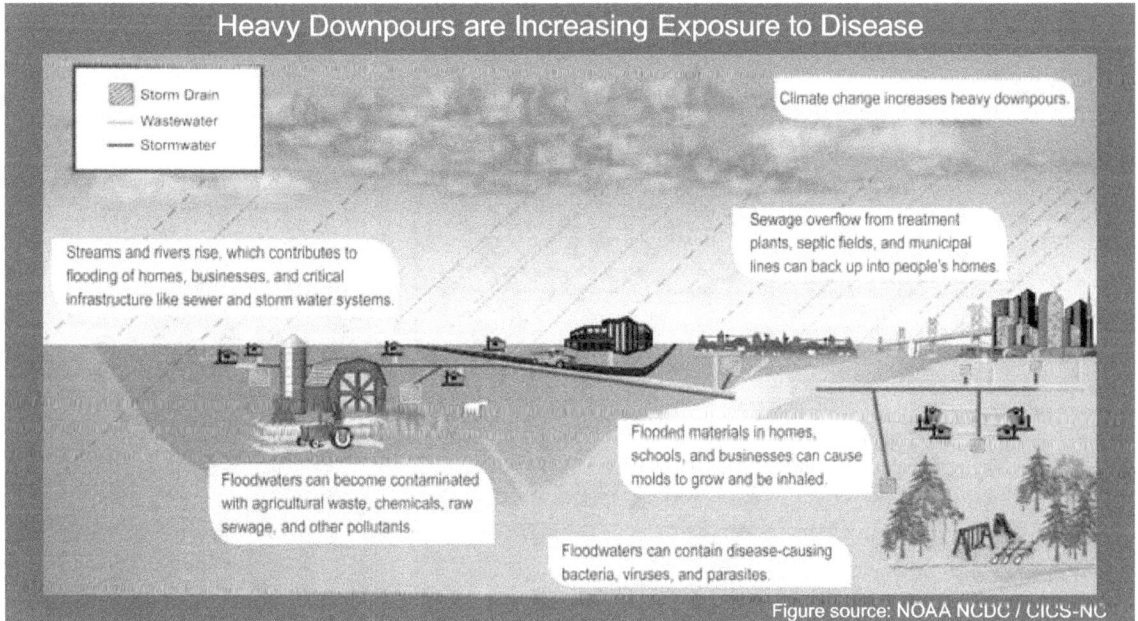

Heavy Downpours are Increasing Exposure to Disease

Storm Drain
Wastewater
Stormwater

Climate change increases heavy downpours.

Streams and rivers rise, which contributes to flooding of homes, businesses, and critical infrastructure like sewer and storm water systems.

Sewage overflow from treatment plants, septic fields, and municipal lines can back up into people's homes.

Floodwaters can become contaminated with agricultural waste, chemicals, raw sewage, and other pollutants.

Flooded materials in homes, schools, and businesses can cause molds to grow and be inhaled.

Floodwaters can contain disease-causing bacteria, viruses, and parasites.

Figure source: NOAA NCDC / CICS-NC

Food and Waterborne Diarrheal Disease

Diarrheal disease is a major public health issue in developing countries and while not generally increasing in the United States, remains a persistent concern nonetheless. Exposure to a variety of pathogens in water and food causes diarrheal disease. Air and water temperatures, precipitation patterns, extreme rainfall events, and seasonal variations are all known to affect disease transmission.[21] In the U.S., children and the elderly are most vulnerable to serious outcomes, and those exposed to inadequately or untreated groundwater will be among those most affected.

In general, diarrheal diseases including Salmonellosis and Campylobacteriosis are more common when temperatures are higher,[22] though patterns differ by place and pathogen. Diarrheal diseases have also been found to occur more frequently in conjunction with both unusually high and low precipitation.[23] Sporadic increases in streamflow rates, often preceded by rapid snowmelt[24] and changes in water treatment,[25] have also been shown to precede outbreaks. Risks of waterborne illness, and beach closures resulting from heavy rain and rising water temperatures are expected to increase in the Great Lakes region due to projected climate change.[26,27]

Extreme Heat

Extreme heat events are the leading weather-related cause of death in the United States.[28] Many cities, including St. Louis, Philadelphia, Chicago, and Cincinnati have suffered dramatic spikes in death rates during heat waves. Deaths result from heat stroke and related conditions,[29] but also from cardiovascular disease, respiratory disease, and cerebrovascular disease.[30,31] Heat waves are also associated with increased hospital admissions for cardiovascular, kidney, and respiratory disorders.[31,32]

Extreme summer heat is increasing in the United States. The effects of heat stress are greatest during heat waves lasting several days or more. As human-induced climate change causes temperatures to continue to rise, heat waves are projected to increase in frequency, intensity, and duration.[33]

Some of the risks of heat-related sickness and death have diminished in recent decades, possibly due to better forecasting, heat-health early warning systems, and/or increased access to air conditioning for the U.S. population.[34] However, extreme heat events remain a cause of preventable death nationwide. Urban heat islands, combined with an aging population and increased urbanization, are projected to increase the vulnerability of urban populations, especially the poor, to heat-related health impacts in the future.[35]

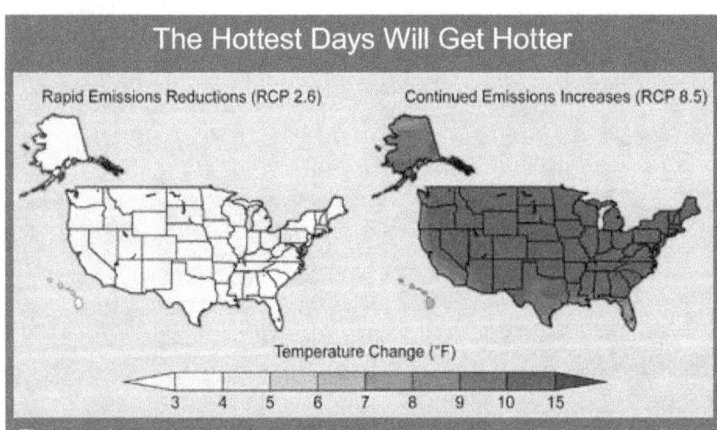

The Hottest Days Will Get Hotter

Rapid Emissions Reductions (RCP 2.6) Continued Emissions Increases (RCP 8.5)

Temperature Change (°F)

3 4 5 6 7 8 9 10 15

The maps show projected increases in the average temperature on the hottest days by late this century (2081-2100) relative to 1986-2005 under a scenario that assumes a rapid reduction in heat-trapping gases (RCP 2.6) and a scenario that assumes continued increases in these gases (RCP 8.5). The hottest days are those so hot they occur only once in 20 years. Across most of the continental U.S., those days will be about 10°F to 15°F hotter in the future under the higher emissions scenario, increasing health risks. (Figure source: NOAA NCDC / CICS-NC).

While deaths and injuries related to extreme cold events are projected to decline due to climate change, these reductions are not expected to compensate for the increase in heat-related deaths.[36]

Climate is one of the factors that influences the distribution of diseases borne by vectors (such as fleas, ticks, and mosquitoes, which spread pathogens that cause illness).[37,38,39,40] The geographic and seasonal distribution of vector populations, and the diseases they can carry, depend not only on climate, but also on land use, socioeconomic and cultural factors, pest control, access to health care, and human responses to disease risk, among other factors.[38,41,42]

North Americans are currently at risk from numerous vector-borne diseases, including Lyme, dengue fever, West Nile virus, Rocky Mountain spotted fever, plague, and tularemia.[40,43,44] Vector-borne pathogens not currently found in the U.S., such as chikungunya, Chagas disease, and Rift Valley fever viruses, are also threats. Climate change effects on the geographical distribution and incidence of vector-borne diseases in other countries where these diseases are already found can also affect North Americans, especially as a result of increasing trade with, and travel to, tropical and subtropical areas.[39,42]

LYME DISEASE

The development and survival of blacklegged ticks, their animal hosts, and the bacterium that causes Lyme disease, are strongly influenced by climatic factors, especially temperature, precipitation, and humidity. Potential impacts of climate change on the transmission of Lyme disease include: 1) changes in the geographic distribution of the disease due to the increase in favorable habitat for ticks to survive off their hosts;[45] 2) a lengthened transmission season due to earlier onset of higher temperatures in the spring and later onset of cold and frost; 3) higher tick densities leading to greater risk in areas where the disease is currently observed due to milder winters and potentially larger rodent host populations; and 4) changes in human behaviors, including increased time outdoors, which may lead to a higher risk of exposure to infected ticks.

Projected Changes in Tick Habitat

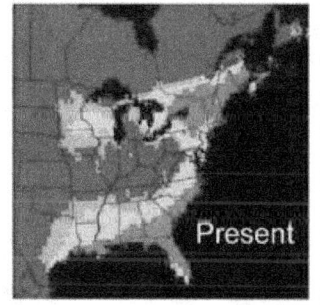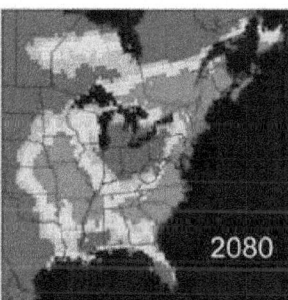

Tick Establishment Probability (%)

0-19 20-39 40-59 60-79 80-99

The maps show the current and projected (for 2080) probability of establishment of tick populations (*Ixodes scapularis*) that transmit Lyme disease. The projected expansion of tick habitat includes much of the eastern half of the country by 2080. For some areas around the Gulf Coast, the probability of tick population establishment is projected to decrease by 2080. (Figure source: adapted from Brownstein et al. 2005[46]).

Multiple Benefits

Policies and other strategies intended to reduce carbon pollution and mitigate climate change can often have independent influences on human health. For example, reducing CO_2 emissions through renewable electrical power generation can reduce air pollutants like particles and sulfur dioxide. Efforts to improve the resiliency of communities and human infrastructure to climate change impacts can also improve human health. There is a growing recognition that the magnitude of health "co-benefits," like reducing both pollution and cardiovascular disease, could be significant, both from a public health and an economic standpoint.[47]

Innovative urban design could create increased access to active transport (such as walking and biking) [27] The compact geographical area found in cities presents opportunities to reduce energy use and emissions of heat-trapping gases and other air pollutants through active transit, improved building construction, provision of services, and infrastructure creation, such as bike paths and sidewalks.[48,49] Urban planning strategies designed to reduce the urban heat island effect, such as green/cool roofs, increased green space, parkland, and urban canopy, could reduce indoor temperatures and improve indoor air quality, and could also produce additional societal co-benefits by promoting social interaction and prioritizing vulnerable urban populations.[48,50]

FINDING
6 INFRASTRUCTURE

Infrastructure is being damaged by sea level rise, heavy downpours, and extreme heat; damages are projected to increase with continued climate change.

Sea level rise, storm surge, and heavy downpours, in combination with the pattern of continued development in coastal areas, are increasing damage to U.S. infrastructure including roads, buildings, and industrial facilities, and are also increasing risks to ports and coastal military installations. Flooding along rivers, lakes, and in cities following heavy downpours, prolonged rains, and rapid melting of snowpack is exceeding the limits of flood protection infrastructure designed for historical conditions. Extreme heat is damaging transportation infrastructure such as roads, rail lines, and airport runways.

Infrastructure around the country has been compromised by extreme weather events and rising sea levels. Power outages and road and bridge damage are among the infrastructure failures that have occurred during these extreme events. A disruption in any one system affects others. For example, a failure of the electrical grid can affect everything from water treatment to public health.

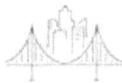

U.S. GLOBAL CHANGE RESEARCH PROGRAM

KEY MESSAGES: URBAN SYSTEMS, INFRASTRUCTURE, AND VULNERABILITY

Climate change and its impacts threaten the well-being of urban residents in all U.S. regions. Essential infrastructure systems such as water, energy supply, and transportation will increasingly be compromised by interrelated climate change impacts. The nation's economy, security, and culture all depend on the resilience of urban infrastructure systems.

In urban settings, climate-related disruptions of services in one infrastructure system will almost always result in disruptions in one or more other infrastructure systems.

Climate vulnerability and adaptive capacity of urban residents and communities are influenced by pronounced social inequalities that reflect age, ethnicity, gender, income, health, and (dis)ability differences.

City government agencies and organizations have started adaptation plans that focus on infrastructure systems and public health. To be successful, these adaptation efforts require cooperative private sector and governmental activities, but institutions face many barriers to implementing coordinated efforts.

Climate change poses a series of interrelated challenges to the country's most densely populated places: its cities. The U.S. is highly urbanized, with about 80% of its population living in cities and metropolitan areas. Cities depend on infrastructure, like water and sewage systems, roads, bridges, and power plants, much of which is aging and in need of repair or replacement. These issues will be compounded by rising sea levels, storm surges, heat waves, and extreme weather events, stressing or even overwhelming essential services.

Urban dwellers are particularly vulnerable to disruptions in essential infrastructure services, in part because many of these infrastructure systems are reliant on each other. For example, electricity is essential to multiple systems, and a failure in the electrical grid can affect water treatment, transportation services, and public health. These infrastructure systems – lifelines to millions – will continue to be affected by various climate-related events and processes.

New York City's subway system, the nation's busiest, sustained the worst damage in its 108 years of operation on October 29, 2012. Millions of people were left without service for at least a week. The damages from Superstorm Sandy are indicative of what powerful tropical storms and higher sea levels could bring more frequently in the future, and were very much in line with vulnerability assessments conducted over the past four years.[4] The effects of the storm would have been far worse if local climate resilience strategies had not been in place. The City of New York and the Metropolitan Transportation Authority worked aggressively to protect life and property by stopping the operation of the city's subway before the storm hit and moving the train cars out of low-lying, flood-prone areas. Catastrophic loss of life would have resulted if there had been subway trains operating in the tunnels when the storm struck.

Cities have become early responders to climate change challenges and opportunities. Integrating climate change action in everyday city and infrastructure operations and governance is an important planning and implementation tool for advancing adaptation in cities.[1,5] By integrating climate-change considerations into daily operations, these efforts can forestall the need to develop a new and isolated set of climate-change-specific policies or procedures.[3] This strategy enables cities and other government agencies to take advantage of existing funding sources and programs, and achieve co-benefits in areas such as sustainability, public health, economic development, disaster preparedness, and environmental justice. Pursuing low-cost, no-regrets options is a particularly attractive short-term strategy for many cities.[1,2]

KEY MESSAGES: TRANSPORTATION

The impacts from sea level rise and storm surge, extreme weather events, higher temperatures and heat waves, precipitation changes, Arctic warming, and other climatic conditions are affecting the reliability and capacity of the U.S. transportation system in many ways.

Sea level rise, coupled with storm surge, will continue to increase the risk of major coastal impacts on transportation infrastructure, including both temporary and permanent flooding of airports, ports and harbors, roads, rail lines, tunnels, and bridges.

Extreme weather events currently disrupt transportation networks in all areas of the country; projections indicate that such disruptions will increase.

Climate change impacts will increase the total costs to the nation's transportation systems and their users, but these impacts can be reduced through rerouting, mode change, and a wide range of adaptive actions.

Transportation systems are affected by climate change and also contribute to climate change. In 2010, the U.S. transportation sector accounted for 27% of all U.S. heat-trapping greenhouse gas emissions, with cars and trucks accounting for 65% of that total.[5] Petroleum accounts for 93% of the nation's transportation energy use.[5] This means that policies and behavioral changes aimed at reducing greenhouse gas emissions will have significant implications for the various components of the transportation sector.

Transportation systems are already experiencing costly climate change related impacts. Many inland states, including Vermont, Tennessee, Iowa, and Missouri, have experienced severe precipitation events, hail, and flooding during the past three years, damaging roads, bridges, and rail systems and the vehicles that use them. Over the coming decades, all modes of transportation and regions will be affected by increasing temperatures, more extreme weather events, and changes in precipitation. Concentrated transportation impacts are particularly expected to occur in Alaska and along seacoasts.

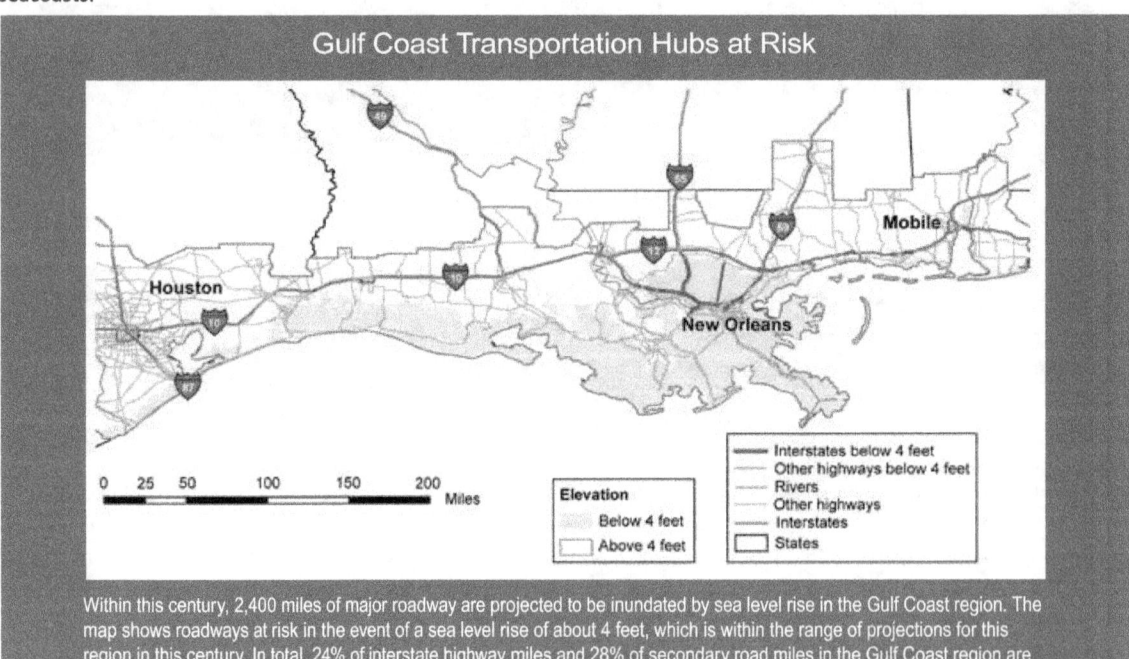

Gulf Coast Transportation Hubs at Risk

Elevation
Below 4 feet
Above 4 feet

Interstates below 4 feet
Other highways below 4 feet
Rivers
Other highways
Interstates
States

Within this century, 2,400 miles of major roadway are projected to be inundated by sea level rise in the Gulf Coast region. The map shows roadways at risk in the event of a sea level rise of about 4 feet, which is within the range of projections for this region in this century. In total, 24% of interstate highway miles and 28% of secondary road miles in the Gulf Coast region are at elevations below 4 feet. (Figure source: Kafalenos et al. 2008[6]).

Key Messages: Energy Supply and Use

Extreme weather events are affecting energy production and delivery facilities, causing supply disruptions of varying lengths and magnitudes and affecting other infrastructure that depends on energy supply. The frequency and intensity of certain types of extreme weather events are expected to change.

Higher summer temperatures will increase electricity use, causing higher summer peak loads, while warmer winters will decrease energy demands for heating. Net electricity use is projected to increase.

Changes in water availability, both episodic and long-lasting, will constrain different forms of energy production.

In the longer term, sea level rise, extreme storm surge events, and high tides will affect coastal facilities and infrastructure on which many energy systems, markets, and consumers depend.

As new investments in energy technologies occur, future energy systems will differ from today's in uncertain ways. Depending on the character of changes in the energy mix, climate change will introduce new risks as well as opportunities.

The U.S. energy system provides a secure supply of energy with only occasional interruptions. However, projected impacts of climate change will increase energy use in the summer and pose additional risks to reliability. Extreme weather events and water shortages are already interrupting energy supply and impacts are expected to increase in the future. Most vulnerabilities and risks to energy supply and use are unique to local situations; others are national in scope.

Increase in Cooling Demand and Decrease in Heating Demand

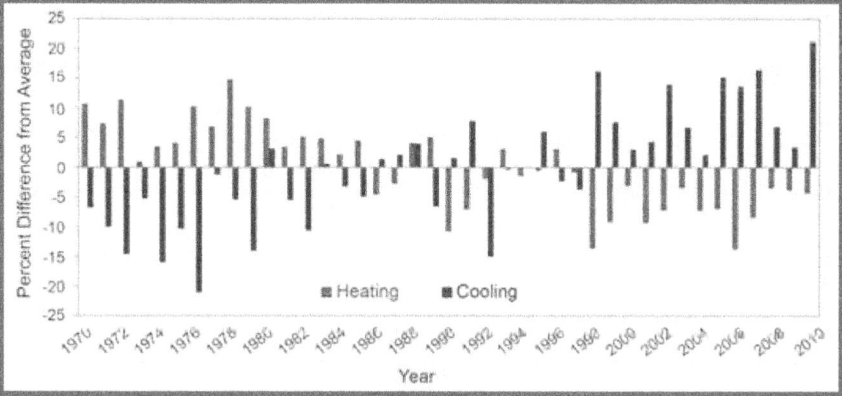

The observed increase in cooling energy demand has been greater than the decrease in heating energy demand. Figure shows observed increases in population-weighted cooling degree days, which result in increased air conditioning use, and decreases in population-weighted heating degree days, meaning less energy required to heat buildings in winter, compared to the average for 1970-2000. Cooling degree days are defined as the number of degrees that a day's average temperature is above 65°F, while heating degree days are the number of degrees a day's average temperature is below 65°F. (Data from NOAA NCDC 2012[8]).

Increases in average temperatures and high temperature extremes are expected to lead to increasing demands for electricity for cooling in every U.S. region. Virtually all cooling load is handled by the electrical grid. In order to meet increased demands for peak electricity, additional generating and distribution facilities will be needed, or demand will have to be managed through a variety of mechanisms. Electricity at peak demand typically is more expensive to supply than at average demand.[7]

In addition to being vulnerable to the effects of climate change, electricity generation is a major source of the heat-trapping gases that contribute to climate change. As a result, regulatory or policy efforts aimed at reducing emissions would also affect the energy supply system.

7 WATER

Water quality and water supply reliability are jeopardized by climate change in a variety of ways that affect ecosystems and livelihoods.

Surface and groundwater supplies in some regions are already stressed by increasing demand as well as declining runoff and groundwater recharge. In some regions, particularly the southern U.S. and the Caribbean and Pacific islands, climate change is increasing the likelihood of water shortages and competition for water. Water quality is diminishing in many areas, particularly due to increasing sediment and contaminant concentrations after heavy downpours.

KEY MESSAGES: WATER RESOURCES

Climate Change Impacts on the Water Cycle

Annual precipitation and river-flow increases are observed now in the Midwest and the Northeast regions. Very heavy precipitation events have increased nationally and are projected to increase in all regions. The length of dry spells is projected to increase in most areas, especially the southern and northwestern portions of the contiguous United States.

Short-term (seasonal or shorter) droughts are expected to intensify in most U.S. regions. Longer-term droughts are expected to intensify in large areas of the Southwest, southern Great Plains, and Southeast.

Flooding may intensify in many U.S. regions, even in areas where total precipitation is projected to decline.

Climate change is expected to affect water demand, groundwater withdrawals, and aquifer recharge, reducing groundwater availability in some areas.

Sea level rise, storms and storm surges, and changes in surface and groundwater use patterns are expected to compromise the sustainability of coastal freshwater aquifers and wetlands.

Increasing air and water temperatures, more intense precipitation and runoff, and intensifying droughts can decrease river and lake water quality in many ways, including increases in sediment, nitrogen, and other pollutant loads.

Climate Change Impacts on Water Resources Use and Management

Climate change affects water demand and the ways water is used within and across regions and economic sectors. The Southwest, Great Plains, and Southeast are particularly vulnerable to changes in water supply and demand.

Changes in precipitation and runoff, combined with changes in consumption and withdrawal, have reduced surface and groundwater supplies in many areas. These trends are expected to continue, increasing the likelihood of water shortages for many uses.

Increasing flooding risk affects human safety and health, property, infrastructure, economies, and ecology in many basins across the United States.

Adaptation and Institutional Responses

In most U.S. regions, water resources managers and planners will encounter new risks, vulnerabilities, and opportunities that may not be properly managed within existing practices.

Increasing resilience and enhancing adaptive capacity provide opportunities to strengthen water resources management and plan for climate change impacts. Many institutional, scientific, economic, and political barriers present challenges to implementing adaptive strategies.

Water Stress in the U. S.

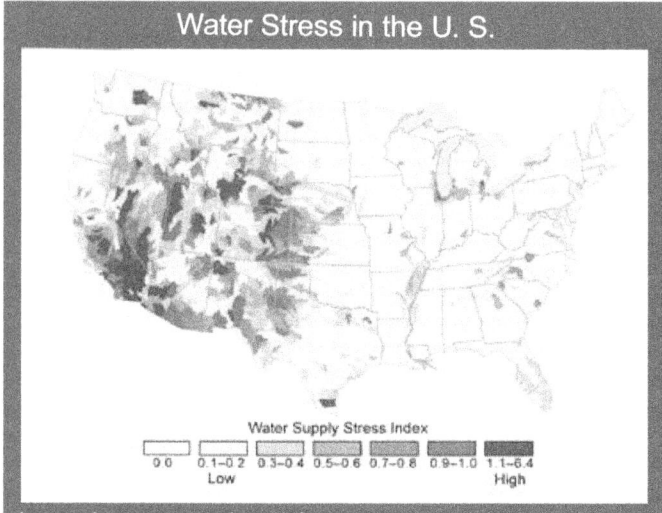

Water Supply Stress Index

| 0 0 | 0.1–0.2 | 0.3–0 4 | 0.5–0 6 | 0.7–0 8 | 0.9–1.0 | 1.1–6.4 |

Low   High

In many places, competing demands for water create stress in local and regional watersheds. Map shows a "water supply stress index" for the U.S. based on observations, with widespread stress in much of the Southwest, western Great Plains, and parts of the Northwest. From an energy production and demand context, watersheds are considered stressed when water demand from agriculture, power plants, and municipalities exceeds 40% of available supply. This often causes conflict for water resources among sectors. In other contexts, many basins experience critical stresses far below this threshold. (Figure source: Averyt et al. 2011[3]).

Changes to Water Demand and Use

Climate change, acting concurrently with demographic, land-use, energy generation and use, and socioeconomic changes, is challenging existing water management practices by affecting water availability and demand and by exacerbating competition among uses and users. In some regions, these current and expected impacts are hastening efficiency improvements in water withdrawal and use, the deployment of more proactive water management and adaptation approaches, and the re-assessment of the water infrastructure and institutional responses.[1]

Water Withdrawals

Total freshwater withdrawals (including water withdrawn and consumed as well as water that returns to the original source) and consumptive uses have leveled off nationally since 1980 at 350 billion gallons of withdrawn water and 100 billion gallons of consumptive water per day, despite the addition of 68 million people from 1980 to 2005.[2] Irrigation and electric power plant cooling withdrawals account for approximately 77% of total withdrawals, municipal and industrial for 20%, and livestock and aquaculture for 3%. Most thermoelectric withdrawals are returned back to rivers after their use for power plant cooling, while most irrigation withdrawals are consumed by the processes of evapotranspiration (evaporation and loss of moisture from leaves) and plant growth. Thus, consumptive water use is dominated by irrigation (81%) followed distantly by municipal and industrial (8%) and the remaining water uses (5%). The largest withdrawals occur in the drier western states for crop irrigation. In the east, water withdrawals mainly serve municipal, industrial, and thermoelectric uses. Some of the largest demand increases are projected in regions where groundwater aquifers are the main water supply source, such as the Great Plains and parts of the Southwest and Southeast. The projected water demand increases (shown below) combined with potentially declining recharge rates threaten the sustainability of many aquifers.

Projected Changes in Water Withdrawals

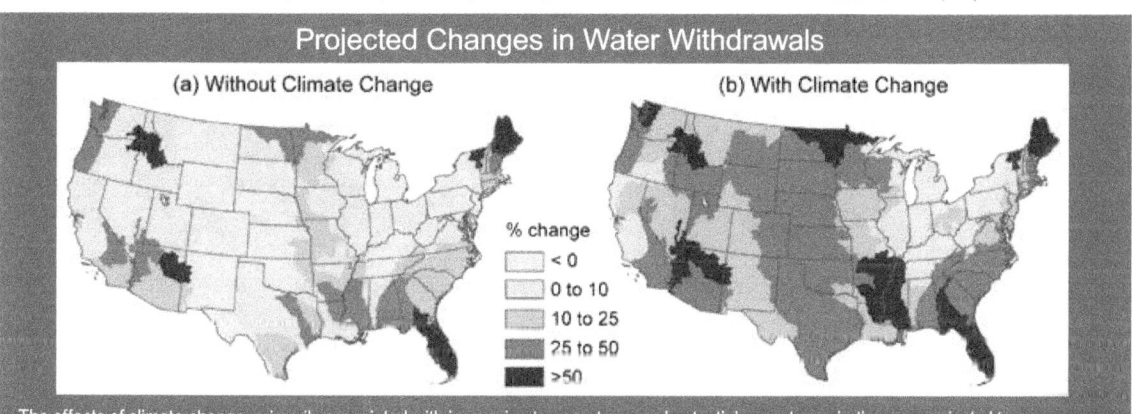

(a) Without Climate Change   (b) With Climate Change

% change
- < 0
- 0 to 10
- 10 to 25
- 25 to 50
- >50

The effects of climate change, primarily associated with increasing temperatures and potential evapotranspiration, are projected to significantly increase water demand across most of the United States. Maps show percent change from 2005 to 2060 in projected demand for water assuming (a) change in population and socioeconomic conditions consistent with the A1B emissions scenario (increasing emissions through the middle of this century, with gradual reductions thereafter), but with no change in climate, and (b) combined changes in population, socioeconomic conditions, and climate according to the A1B emissions scenario. (Figure source: Brown et al. 2013[4])

Projected Snow Water Equivalent

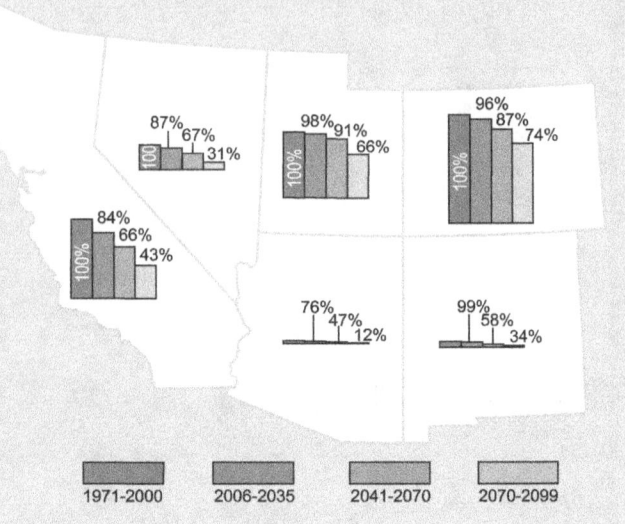

Snow water equivalent refers to the amount of water held in a volume of snow, which depends on the density of the snow and other factors. Figure shows projected snow water equivalent for the Southwest, as a percentage of 1971-2000 levels, assuming continued increases in global emissions (A2 scenario). The size of the bars is in proportion to the amount of snow each state contributes to the regional total; thus, the bars for Arizona are much smaller than those for Colorado, which contributes the most to region-wide snowpack. Declines in peak snow water equivalent are strongly correlated with early timing of runoff and decreases in total runoff. For watersheds that depend on snowpack to provide the majority of the annual runoff, such as in the Sierra Nevada and in the Upper Colorado and Upper Rio Grande River Basins, lower snow water equivalent generally translates to reduced reservoir water storage. (Data from Scripps Institution of Oceanography).

1971-2000   2006-2035   2041-2070   2070-2099

Water Quality

Lower and more persistent low flows under drought conditions as well as higher flows during floods can worsen water quality. Increasing precipitation intensity, along with the effects of wildfires and fertilizer use, are increasing sediment, nutrient, and contaminant loads in surface waters used by downstream water users[5] and ecosystems in some places. Changing land cover, flood frequencies, and flood magnitudes are expected to increase mobilization of sediments in large river basins.[6]

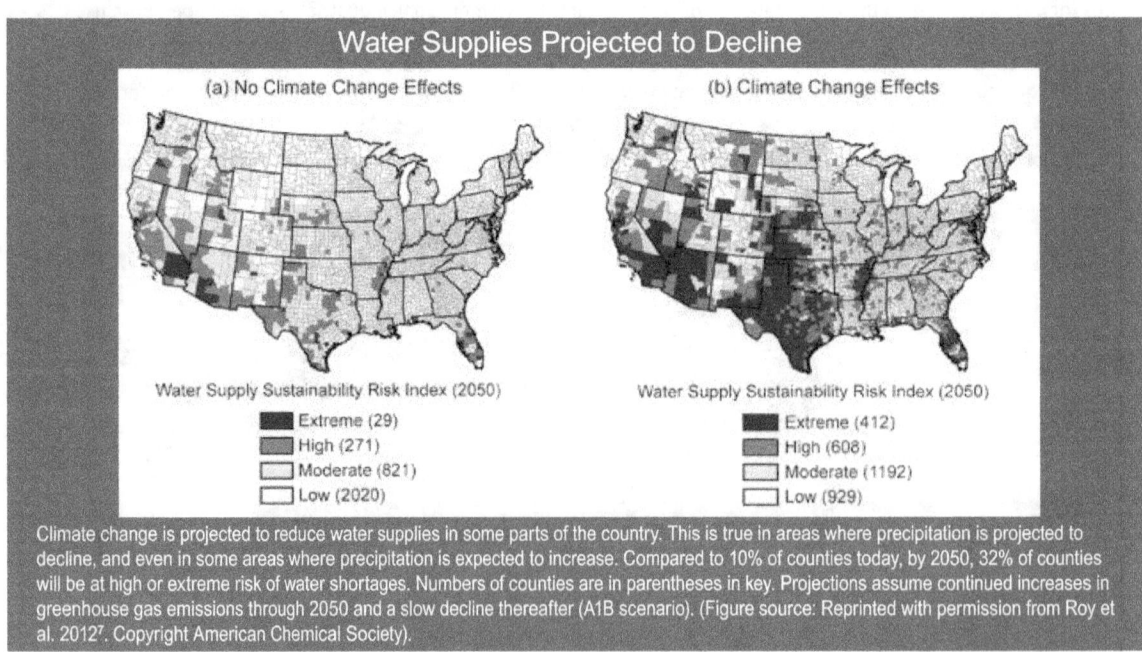

Water Supplies Projected to Decline

(a) No Climate Change Effects

Water Supply Sustainability Risk Index (2050)

Extreme (29)
High (271)
Moderate (821)
Low (2020)

(b) Climate Change Effects

Water Supply Sustainability Risk Index (2050)

Extreme (412)
High (608)
Moderate (1192)
Low (929)

Climate change is projected to reduce water supplies in some parts of the country. This is true in areas where precipitation is projected to decline, and even in some areas where precipitation is expected to increase. Compared to 10% of counties today, by 2050, 32% of counties will be at high or extreme risk of water shortages. Numbers of counties are in parentheses in key. Projections assume continued increases in greenhouse gas emissions through 2050 and a slow decline thereafter (A1B scenario). (Figure source: Reprinted with permission from Roy et al. 2012[7]. Copyright American Chemical Society).

KEY MESSAGES: ENERGY, WATER, AND LAND USE

Energy, water, and land systems interact in many ways. Climate change affects the individual sectors and their interactions; the combination of these factors affects climate change vulnerability as well as adaptation and mitigation options for different regions of the country.

The dependence of energy systems on land and water supplies will influence the development of these systems and options for reducing greenhouse gas emissions, as well as their climate change vulnerability.

Jointly considering risks, vulnerabilities, and opportunities associated with energy, water, and land use is challenging, but can improve the identification and evaluation of options for reducing climate change impacts.

Energy production, land use, and water resources are linked in complex ways. Electric utilities and energy companies compete with farmers and ranchers for water rights in some parts of the country. Land-use planners need to consider the interactive impacts of strained water supplies on cities, agriculture, and ecological needs. Across the country, these intertwined sectors will witness increased stresses due to climate changes that are projected to reduce water quality and/or quantity in many regions and change heating and cooling electricity demand, among other impacts.

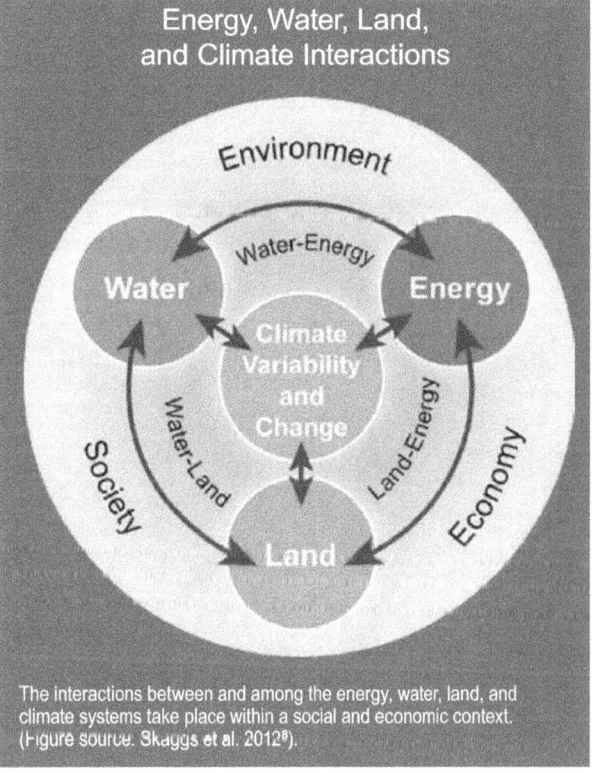

The interactions between and among the energy, water, land, and climate systems take place within a social and economic context. (Figure source: Skaggs et al. 2012[8]).

FINDING
8 AGRICULTURE

Climate disruptions to agriculture have been increasing and are projected to become more severe over this century.

Some areas are already experiencing climate-related disruptions, particularly due to extreme weather events. While some U.S. regions and some types of agricultural production will be relatively resilient to climate change over the next 25 years or so, others will increasingly suffer from stresses due to extreme heat, drought, disease, and heavy downpours. From mid-century on, climate change is projected to have more negative impacts on crops and livestock across the country – a trend that could diminish the security of our food supply.

KEY MESSAGES: AGRICULTURE

Climate disruptions to agricultural production have increased in the past 40 years and are projected to increase over the next 25 years. By mid-century and beyond, these impacts will be increasingly negative on most crops and livestock.

Many agricultural regions will experience declines in crop and livestock production from increased stress due to weeds, diseases, insect pests, and other climate change induced stresses.

Current loss and degradation of critical agricultural soil and water assets due to increasing extremes in precipitation will continue to challenge both rainfed and irrigated agriculture unless innovative conservation methods are implemented.

The rising incidence of weather extremes will have increasingly negative impacts on crop and livestock productivity because critical thresholds are already being exceeded.

Agriculture has been able to adapt to recent changes in climate; however, increased innovation will be needed to ensure the rate of adaptation of agriculture and the associated socioeconomic system can keep pace with climate change over the next 25 years.

Climate change effects on agriculture will have consequences for food security, both in the U.S. and globally, through changes in crop yields and food prices and effects on food processing, storage, transportation, and retailing. Adaptation measures can help delay and reduce some of these impacts.

Crop Yields Decline under Higher Temperatures

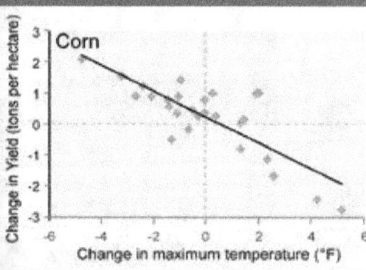
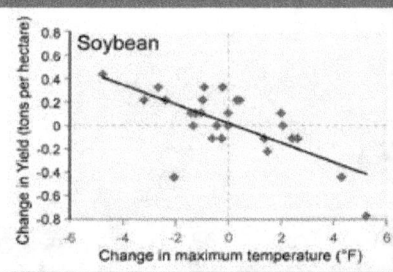

Crop yields are very sensitive to temperature and rainfall. They are especially sensitive to high temperatures during the pollination and grain-filling period. For example, corn (left) and soybean (right) harvests in Illinois and Indiana, two major producers, were lower in years with average maximum summer (June, July, and August) temperatures that were higher than the 1980-2007 average. Most years with below-average yields are both warmer and drier than normal.[1,2] There is a very high correlation between warm and dry conditions during Midwest summers[3] due to similar meteorological conditions and drought-caused changes[4] in the land surface. (Figure source: redrawn from Mishra and Cherkauer 2010[1]).

Climate change poses a major challenge to U.S. agriculture, because of the critical dependence of the agricultural system on climate and because of the complex role agriculture plays in social and economic systems. Climate change has the potential to both positively and negatively affect the location, timing, and productivity of crop, livestock, and fishery systems at local, national, and global scales.

The U.S. produces nearly $330 billion per year in agricultural commodities.[5] This productivity is vulnerable to direct impacts on crop and livestock development and yield from changing climate conditions and extreme weather events, and indirect impacts through increasing pressures from pests and pathogens. Climate change has the potential to both positively and negatively affect agricultural systems at local, national, and global scales. Climate change will also alter the stability of food supplies and create new food security challenges for the U.S. as the world seeks to feed nine billion people by 2050.

The agricultural sector continually adapts through a variety of strategies that have allowed previous agricultural production to increase, as evidenced by the continued growth in production and efficiency across the United States. However, the magnitude of climate change projected for this century and beyond, particularly under higher emissions scenarios, will challenge the ability of the agriculture sector to continue to successfully adapt.

Key Climate Variables Affecting Agricultural Productivity

Change in Frost-Free Season Length

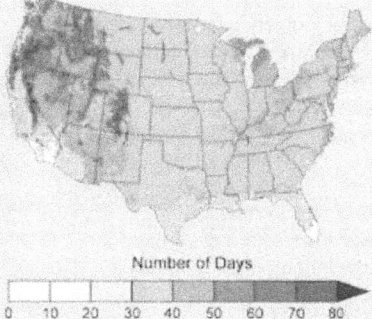

Number of Days

Frost-free season is projected to lengthen across much of the nation. Taking advantage of the increasing length of the growing season and changing planting dates could allow planting of more diverse crop rotations, which can be an effective adaptation strategy.

Change in Number of Consecutive Dry Days

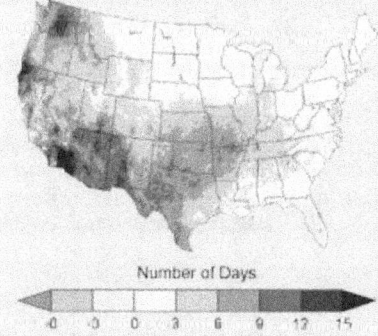

Number of Days

The annual maximum number of consecutive dry days (less than 0.01 inches of rain) is projected to increase, especially in the western and southern parts of the nation, negatively affecting crop and animal production. The trend toward more consecutive dry days and higher temperatures will increase evaporation and add stress to limited water resources, affecting irrigation and other water uses.[6]

Change in Number of Hot Nights

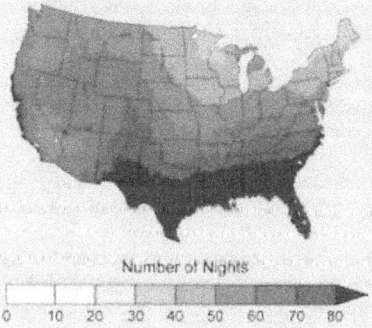

Number of Nights

Hot nights are defined as nights with a minimum temperature higher than 98% of the minimum temperatures between 1971 and 2000. Such nights are projected to increase throughout the nation. High nighttime temperatures can reduce grain yields and increase stress on animals, resulting in reduced rates of meat, milk, and egg production.[7]

Projections are shown for 2070-2099 as compared to 1971-2000 under an emissions scenario that assumes continued increases in heat-trapping gases (A2). (Figure source: NOAA NODC / CICS NC).

9 INDIGENOUS PEOPLES

Climate change poses particular threats to Indigenous Peoples' health, well-being, and ways of life.

The peoples, lands, and resources of indigenous communities in the United States, including Alaska and the Pacific Rim, face an array of climate change impacts and vulnerabilities. The consequences of observed and projected climate change have and will undermine indigenous ways of life that have persisted for thousands of years. Native cultures are directly tied to Native places and homelands, and many indigenous peoples regard all people, plants, and animals that share our world as relatives rather than resources. Language, ceremonies, cultures, practices, and food sources evolved in concert with the inhabitants, human and non-human, of specific homelands.

Human-caused stresses such as dam building have greatly reduced salmon on the Klamath River.

Climate change impacts on many of the 566 federally recognized tribes and other tribal and indigenous groups are projected to be especially severe, since these impacts are compounded by a number of persistent social and economic problems.[1] Key vulnerabilities include the loss of traditional knowledge in the face of rapidly changing ecological conditions, increased food insecurity due to reduced availability of traditional foods, changing water availability, Arctic sea ice loss, permafrost thaw, and relocation from historic homelands.[2,3]

We humbly ask permission from all our relatives; our elders, our families, our children, the winged and the insects, the four-legged, the swimmers, and all the plant and animal nations, to speak. Our Mother has cried out to us. She is in pain. We are called to answer her cries. Msit No'Kmaq – All my relations!
— Indigenous Prayer

KEY MESSAGES: INDIGENOUS PEOPLES, LANDS, AND RESOURCES

Observed and future impacts from climate change threaten Native Peoples' access to traditional foods such as fish, game, and wild and cultivated crops, which have provided sustenance as well as cultural, economic, medicinal, and community health for generations.

A significant decrease in water quality and quantity due to a variety of factors, including climate change, is affecting drinking water, food, and cultures. Native communities' vulnerabilities and limited capacity to adapt to water-related challenges are exacerbated by historical and contemporary government policies and poor socioeconomic conditions.

Declining sea ice in Alaska is causing significant impacts to Native communities, including increasingly risky travel and hunting conditions, damage and loss to settlements, food insecurity, and socioeconomic and health impacts from loss of cultures, traditional knowledge, and homelands.

Alaska Native communities are increasingly exposed to health and livelihood hazards from increasing temperatures and thawing permafrost, which are damaging critical infrastructure, adding to other stressors on traditional lifestyles.

Climate change related impacts are forcing relocation of tribal and indigenous communities, especially in coastal locations. These relocations, and the lack of governance mechanisms or funding to support them, are causing loss of community and culture, health impacts, and economic decline, further exacerbating tribal impoverishment.

Indigenous communities in various parts of the U.S. have observed climatic changes that result in impacts such as the loss of traditional foods, medicines, and water supplies. The Southwest's 182 federally recognized tribes and communities in its U.S.-Mexico border region share particularly high vulnerabilities to climate changes such as high temperatures, drought, and severe storms. Changes in long-term average temperature, precipitation, and declining snowpack have altered the physical and hydrologic environment on the Colorado Plateau, making the Navajo Nation more susceptible to drought impacts.[4] Southwest tribes have observed damage to agriculture and livestock, the loss of springs and medicinal and culturally important plants and animals, and impacts on drinking water supplies.[5] In the Northwest, tribal treaty rights are being affected by the reduction of rainfall and snowmelt in the mountains, melting glaciers, rising temperatures, and shifts in ocean currents.[6] Tribal communities in coastal Louisiana are experiencing climate change induced rising sea levels, along with saltwater intrusion, subsidence, and intense erosion and land loss due to oil and gas extraction, levees, dams, and other river management techniques, forcing them to either relocate or try to find ways to save their land.[7] In Hawai'I, Native peoples have observed a shortening of the rainy season, increasing intensity of storms and flooding, and unpredictable rainfall patterns.[8]

Harvesting traditional foods is important to Native Peoples' culture, health, and economic well being. In the Great Lakes region, wild rice is unable to grow in its traditional range due to warming winters and changing water levels.

Alaska Natives Face Multiple Climate Impacts

Alaska is home to 40% (229 of 566) of the federally recognized tribes in the United States.[9] The small number of jobs, high cost of living, and rapid social change make rural, predominantly Native, communities highly vulnerable to climate change through impacts on traditional hunting and fishing practices. In Alaska, water availability, quality, and quantity are threatened by the consequences of permafrost thaw, which has damaged community water infrastructure, as well as by the northward extension of diseases such as those caused by the *Giardia* parasite.[10]

Rising temperatures are causing damage in Native villages in Alaska as sea ice declines and permafrost thaws. Resident of Selawik, Alaska, and his granddaughter survey a water line sinking into the thawing permafrost, August 2011.

Arctic regional temperatures have risen at twice the global rate over the past few decades.[2] This temperature increase – which is expected to continue with future climate change – is accompanied by significant reductions in sea ice thickness and extent, increased permafrost thaw, more extreme weather and severe storms, and changes in seasonal ice melt/ freeze of lakes and rivers, water temperature, sea level, flooding patterns, erosion, and snowfall timing and type.[11,12] These changes increase the number of serious problems for Alaska Native populations, which include: injury from extreme or unpredictable weather and thinning sea ice; changing snow and ice conditions that limit safe hunting, fishing, or herding practices; malnutrition and food insecurity from lack of access to subsistence food; contamination of food and water; increasing economic, mental, and social problems from loss of culture and traditional livelihood; increases in infectious diseases; and loss of buildings and infrastructure from permafrost erosion and thawing, resulting in the relocation of entire communities.[2,10,12,13] For more, see pages 82-83.

10 ECOSYSTEMS

Ecosystems and the benefits they provide to society are being affected by climate change. The capacity of ecosystems to buffer the impacts of extreme events like fires, floods, and severe storms is being overwhelmed.

Climate change impacts on biodiversity are already being observed in alteration of the timing of critical biological events such as spring bud burst, and substantial range shifts of many species. In the longer term, there is an increased risk of species extinction. These changes have social, cultural, and economic effects. Events such as droughts, floods, wildfires, and pest outbreaks associated with climate change (for example, bark beetles in the West) are already disrupting ecosystems. These changes limit the capacity of ecosystems, such as forests, barrier beaches, and wetlands, to continue to play important roles in reducing the impacts of extreme events on infrastructure, human communities, and other valued resources.

In addition to direct impacts on ecosystems, societal choices about land use and agricultural practices affect the cycling of carbon, nitrogen, phosphorus, sulfur, and other elements, which also influence climate. These choices can affect, positively or negatively, the rate and magnitude of climate change and the vulnerabilities of human and natural systems.

Changes in snowmelt patterns are affecting water supply. Mt. Rainier, Washington.

KEY MESSAGES: ECOSYSTEMS AND BIODIVERSITY

Climate change impacts on ecosystems reduce their ability to improve water quality and regulate water flows.

Climate change, combined with other stressors, is overwhelming the capacity of ecosystems to buffer the impacts from extreme events like fires, floods, and storms.

Landscapes and seascapes are changing rapidly, and species, including many iconic species, may disappear from regions where they have been prevalent, or become extinct, altering some regions so much that their mix of plant and animal life will become almost unrecognizable.

Timing of critical biological events, such as spring bud burst, emergence from overwintering, and the start of migrations, has shifted, leading to important impacts on species and habitats.

Whole system management is often more effective than focusing on one species at a time, and can help reduce the harm to wildlife, natural assets, and human well-being that climate disruption might cause.

Climate change affects the living world, including people, through changes in ecosystems, biodiversity, and ecosystem services. Ecosystems entail all the living things in a particular area as well as the non-living things with which they interact, such as air, soil, water, and sunlight. Biodiversity refers to the variety of life, including the number of species, life forms, genetic types, and habitats and biomes (which are characteristic groupings of plant and animal species found in a particular climate). Biodiversity and ecosystems produce a rich array of benefits that people depend on, including fisheries, drinking water, fertile soils for growing crops, climate regulation, inspiration, and aesthetic and cultural values.[1] These benefits are called "ecosystem services" — some of which, like food, are more easily quantified than others, such as climate regulation or cultural values. Changes in many such services are often not obvious to those who depend on them.

Ecosystem services contribute to jobs, economic growth, health, and human well-being. Although we interact with ecosystems and ecosystem services every day, their linkage to climate change can be elusive because they are influenced by so many additional entangled factors.[2] Ecosystem perturbations driven by climate change have direct human impacts, including reduced water supply and quality, the loss of iconic species and landscapes, distorted rhythms of nature, and the potential for extreme events to overwhelm the regulating services of ecosystems.

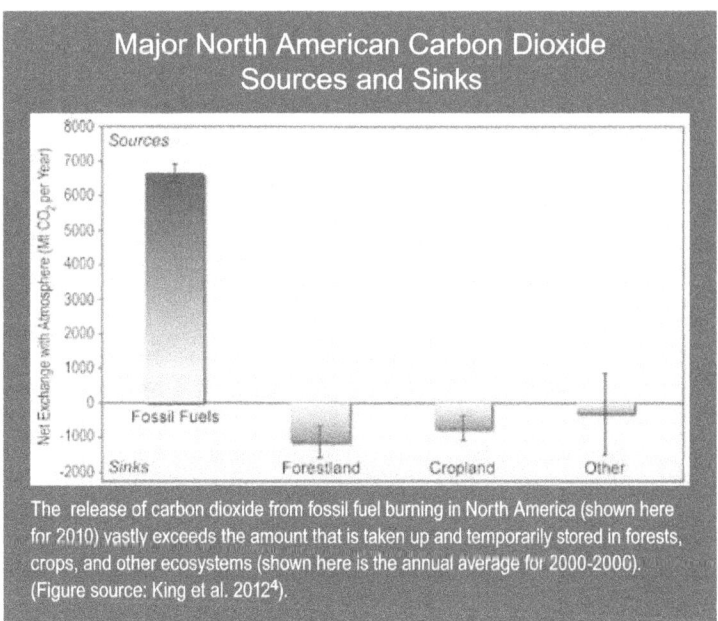

Major North American Carbon Dioxide Sources and Sinks

The release of carbon dioxide from fossil fuel burning in North America (shown here for 2010) vastly exceeds the amount that is taken up and temporarily stored in forests, crops, and other ecosystems (shown here is the annual average for 2000-2006). (Figure source: King et al. 2012[4]).

Even with these well-documented ecosystem impacts, it is often difficult to quantify human vulnerability that results from shifts in ecosystem processes and services. For example, although it is relatively straightforward to predict how precipitation will change water flow, it is much harder to pinpoint which farms, cities, and habitats will be at risk of running out of water, and even more difficult to say how people will be affected by the loss of a favorite fishing spot or a wildflower that no longer blooms in the region. A better understanding of how a range of ecosystem responses affects people – from altered water flows to the loss of wildflowers – will help to inform the management of ecosystems in a way that promotes resilience to climate change.

Ecosystems also represent potential "sinks" for CO_2, which are places where carbon can be stored over the short or long term. At the continental scale, there has been a large and relatively consistent increase in forest carbon stocks over the last two decades,[3] due to recovery from past forest harvest, net increases in forest area, improved forest management regimes, and faster growth driven by climate or fertilization by CO_2 and nitrogen.[4,5] Emissions of CO_2 from human activities in the United States continue to exceed ecosystem CO_2 uptake by more than three times. As a result, North America remains a net source of CO_2 into the atmosphere[4] by a substantial margin.

Forests absorb carbon dioxide and provide many other ecosystem services, such as purifying water and providing recreational opportunities.

KEY MESSAGES: FORESTS

Climate change is increasing the vulnerability of many forests to ecosystem changes and tree mortality through fire, insect infestations, drought, and disease outbreaks.

U.S. forests and associated wood products currently absorb and store the equivalent of about 16% of all carbon dioxide (CO_2) emitted by fossil fuel burning in the U.S. each year. Climate change, combined with current societal trends in land use and forest management, is projected to reduce this rate of forest CO_2 uptake.

Bioenergy could emerge as a new market for wood and could aid in the restoration of forests killed by drought, insects, and fire.

Forest management responses to climate change will be influenced by the changing nature of private forestland owner-ship, globalization of forestry markets, emerging markets for bioenergy, and U.S. climate change policy.

Forests occur within urban areas, at the interface between urban and rural areas (wildland-urban interface), and in rural areas. Urban forests contribute to clean air, cooling buildings, aesthetics, and recreation in parks. Development in the wildland-urban interface is increasing because of the appeal of owning homes near or in the woods. In rural areas, market factors drive land uses among commercial forestry and land uses such as agriculture. Across this spectrum, forests provide recreational opportunities, cultural resources, and social values such as aesthetics.[6]

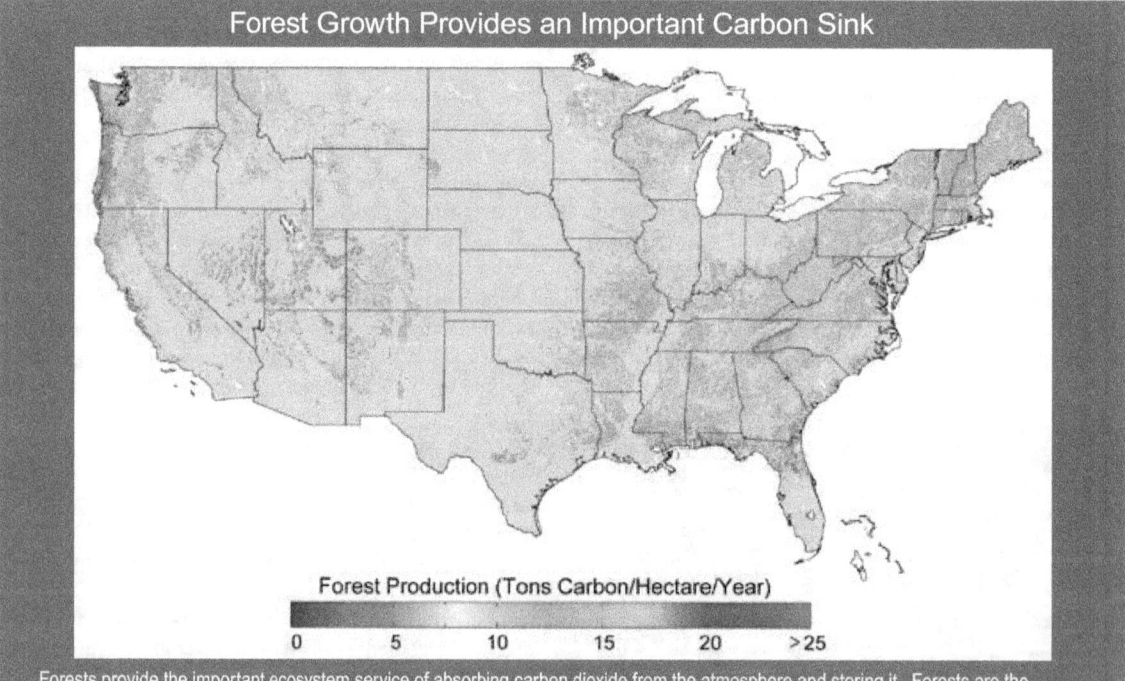

Forest Growth Provides an Important Carbon Sink

Forest Production (Tons Carbon/Hectare/Year)

0   5   10   15   20   >25

Forests provide the important ecosystem service of absorbing carbon dioxide from the atmosphere and storing it. Forests are the largest component of the U.S. carbon sink, but growth rates of forests vary widely across the country. Well-watered forests of the Pacific Coast and Southeast absorb considerably more than the arid Southwestern forests or the colder Northeastern forests. Climate change and disturbance rates, combined with current societal trends regarding land use and forest management, are projected to reduce forest CO_2 uptake in the coming decades. Figure shows forest growth as measured by net primary production in tons of carbon per hectare per year, and are averages from 2000 to 2006 (Figure source: adapted from Running et al. 2004[7]).

Economic factors have historically influenced both the overall area and use of private forestland. Private entities own 56% of U.S. forestlands while 44% of forests are on public lands.[8] Market factors can influence management objectives for public lands, but societal values also influence objectives by identifying benefits such as environmental services not ordinarily provided through markets, like watershed protection and wildlife habitat. Different challenges and opportunities exist for public and for private forest management decisions, especially when climate-related issues are considered on a national scale. For example, public forests typically carry higher levels of forest biomass, are more remote, and tend not to be as intensively managed as private forestlands.[6]

Forests provide opportunities to reduce future climate change by capturing and storing carbon, as well as by providing resources for bioenergy production (the use of forest-derived plant-based materials for energy production). The total amount of carbon stored in U.S. forest ecosystems and wood products (such as lumber and pulpwood) equals roughly 25 years of U.S. heat-trapping gas emissions at current rates of emission, providing an important national "sink" that could grow or shrink depending on the extent of climate change, forest management practices, policy decisions, and other factors.[9]

Climate change is increasing vulnerability to wildfires across the western U.S. and Alaska.

FOREST DISTURBANCE

Factors affecting tree death, such as drought, physiological water stress, higher temperatures, and/or pests and pathogens, are often interrelated, which means that isolating a single cause of mortality is rare.[10] However, in western forests there have been recent large scale die-off events due to one or more of these factors,[11,12,13] and rates of tree mortality are well correlated with both rising temperatures and associated increases in evaporative water demand.[14]

Fire is another important forest disturbance. Given strong relationships between climate and fire, even when modified by land use and management, such as fuel treatments, projected climate changes suggest that western forests in the U.S. will be increasingly affected by large and intense fires that occur more frequently.[13,15]

A Montana saw mill owner inspects a lodgepole pine covered in pitch tubes that show the tree trying, unsuccessfully, to defend itself against the bark beetle. The bark beetle is killing lodgepole pines throughout the western United States.

Warmer winters allow more insects to survive the cold season, and a longer summer allows some insects to complete two life cycles in a year instead of one. Drought stress reduces trees' ability to defend against boring insects. Above, beetle-killed trees in Rocky Mountain National Park in Colorado.

KEY MESSAGES:
LAND USE AND LAND COVER CHANGE

Choices about land-use and land-cover patterns have affected and will continue to affect how vulnerable or resilient human communities and ecosystems are to the effects of climate change.

Land-use and land-cover changes affect local, regional, and global climate processes.

Individuals, businesses, non-profits, and governments have the capacity to make land-use decisions to adapt to the effects of climate change.

Choices about land use and land management may provide a means of reducing atmospheric greenhouse gas levels.

Land-use and land-cover changes affect climate processes. Above, development along Colorado's Front Range.

In addition to emissions of heat-trapping greenhouse gases from energy, industrial, agricultural, and other activities, humans affect climate through changes in land use (activities taking place on land, like growing food, cutting trees, or building cities) and land cover (the physical characteristics of the land surface, including grain crops, trees, or concrete). For example, cities are warmer than the surrounding countryside because the greater extent of paved areas in cities affects how water and energy are exchanged between the land and the atmosphere, and how exposed the population is to extreme heat events. Decisions about land use and land cover can therefore affect, positively or negatively, how much our climate will change, and what kind of vulnerabilities humans and natural systems will face as a result.

The combination of residential location choices with wildfire occurrence dramatically illustrates how the interactions between land use and climate processes can affect climate change impacts and vulnerabilities. Low-density (suburban and exurban) housing patterns in the U.S. have expanded, and are projected to continue to expand.[16] One result is a rise in the amount of construction in forests and other wildlands[17] that in turn has increased the exposure of houses, other structures, and people to damages from wildfires. The number of buildings lost in the 25 most destructive fires in California history increased significantly in the 1990s and 2000s compared to the previous three decades, as shown in the figure.[18] These losses are one example of how changing development patterns can interact with a changing climate to create dramatic new risks. In the western U.S., increasing frequencies of large wildfires and longer wildfire durations are strongly associated with increased spring and summer temperatures and an earlier spring snowmelt.[19]

Construction near forests and wildlands is growing. Here, wildfire approaches a housing development.

Building Loss by Fires at California Wildland-Urban Interfaces

Many forested areas in the U.S. have experienced a recent building boom in what is known as the "wildland-urban interface." This figure shows the number of buildings lost from the 25 most destructive wildland-urban interface fires in California history from 1960 to 2007 (Figure source: Stephens et al. 2009[18]).

54

KEY MESSAGES: BIOGEOCHEMICAL CYCLES

Human activities have increased atmospheric carbon dioxide by about 40% over pre-industrial levels and more than doubled the amount of nitrogen available to ecosystems. Similar trends have been observed for phosphorus and other elements, and these changes have major consequences for biogeochemical cycles and climate change.

In total, land in the U.S. absorbs and stores an amount of carbon equivalent to about 17% of annual U.S. fossil fuel emissions. U.S. forests and associated wood products account for most of this land sink. The effect of this carbon storage is to partially offset warming from emissions of CO_2 and other greenhouse gases.

Altered biogeochemical cycles together with climate change increase the vulnerability of biodiversity, food security, human health, and water quality to changing climate. However, natural and managed shifts in major biogeochemical cycles can help limit rates of climate change.

Biogeochemical cycles involve the fluxes of chemical elements among different parts of the Earth: from living to non-living, from atmosphere to land to sea, and from soils to plants. Human activities have mobilized Earth elements and accelerated their cycles – for example, more than doubling the amount of reactive nitrogen that has been added to the biosphere since pre-industrial times.[20]

Global-scale alterations of biogeochemical cycles are occurring from human activities, both in the U.S. and elsewhere, with impacts and implications now and into the future. Global carbon dioxide emissions are the most significant driver of human-caused climate change. But human-accelerated cycles of other elements, especially nitrogen, phosphorus, and sulfur, also influence climate. These elements can affect climate directly and indirectly, amplifying or reducing the impacts of climate change. Climate change is having, and will continue to have, impacts on biogeochemical cycles, which will alter future impacts on climate and affect our capacity to cope with coupled changes in climate, biogeochemistry, and other factors.

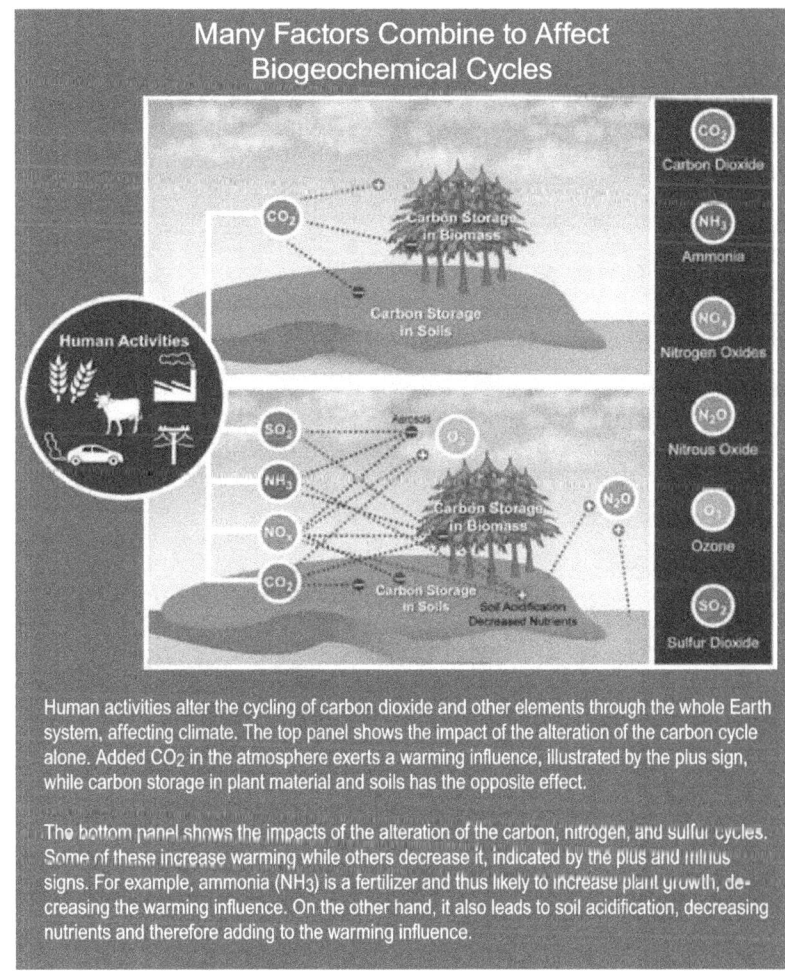

Human activities alter the cycling of carbon dioxide and other elements through the whole Earth system, affecting climate. The top panel shows the impact of the alteration of the carbon cycle alone. Added CO_2 in the atmosphere exerts a warming influence, illustrated by the plus sign, while carbon storage in plant material and soils has the opposite effect.

The bottom panel shows the impacts of the alteration of the carbon, nitrogen, and sulfur cycles. Some of these increase warming while others decrease it, indicated by the plus and minus signs. For example, ammonia (NH_3) is a fertilizer and thus likely to increase plant growth, decreasing the warming influence. On the other hand, it also leads to soil acidification, decreasing nutrients and therefore adding to the warming influence.

SPECIES RESPONSES

Mussel and barnacle beds have declined or disappeared along parts of the Northwest coast due to higher temperatures and drier conditions.[21]

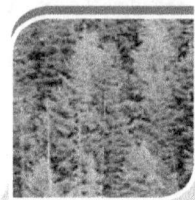

Conifers in many western forests have died, experiencing mortality rates up to 87%, from warming-induced changes in the prevalence of pests and pathogens and drought stress.[12]

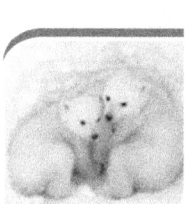

In response to climate-related habitat change, many small mammal species have altered their ranges, with lower-elevation species expanding their ranges and higher-elevation species contracting their ranges.[23]

Decreases in the weight and survival of polar bear offspring along the north Alaska coast have been linked to changes in mother's body size and/or condition following years with lower availability of optimal sea ice habitat.[22]

Quaking aspen tree dominated systems are experiencing declines in the western U.S. due to drought stress during the last decade.[24]

Warmer springs in Alaska have reduced calving success in caribou populations as a result of earlier onset of plant emergence and decreased spatial variation in growth and availability of forage to breeding caribou.[25]

Climate change is likely to influence elevational patterns in vegetation as Hawaiian mountain vegetation types vary in their sensitivity to changes in moisture availability.[26]

Warming-induced interbreeding was detected between southern and northern flying squirrels in the Great Lakes region of Ontario, Canada, and Pennsylvania after a series of warm winters created more overlap in their habitat ranges.[27]

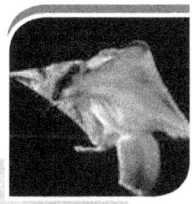

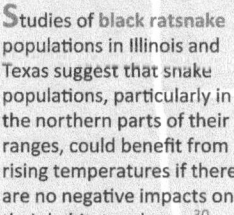

First flowering dates plant species in North Dakota have shifted significantly in more than 40% of the 178 species examined, with the greatest changes observed during the two warmest years of the study.[28]

In the Northwest Atlantic, 24 out of 36 commercial fish stocks showed significant range shifts, both in latitude and depth, between 1968 and 2007 in response to increased sea surface and bottom temperatures.[29]

Studies of black ratsnake populations in Illinois and Texas suggest that snake populations, particularly in the northern parts of their ranges, could benefit from rising temperatures if there are no negative impacts on their habitat and prey.[30]

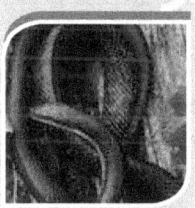

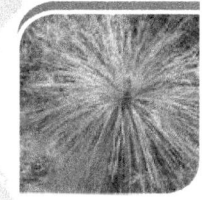

Widespread declines in body size of resident and migrant birds in western Pennsylvania were documented over a 40-year period. The higher the average regional temperatures in the preceding year, the smaller the birds.[31]

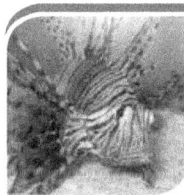

Seedling survival for nearly 20 species of trees decreased during years of lower rainfall in the Southern Appalachians and the Piedmont areas, indicating reductions in native species.[33]

Climatic fluctuations increase the probability of infidelity in birds that are normally monogamous. This increases gene exchange and the likelihood of offspring survival.[32]

Some warm-water fishes have moved northwards, and some tropical and subtropical fishes in the northern Gulf of Mexico have increased in temperate ocean habitat.[34] Similar shifts and invasions have been documented in Long Island Sound and Narragansett Bay in the Atlantic Ocean.[35]

11 OCEANS

Ocean waters are becoming warmer and more acidic, broadly affecting ocean circulation, chemistry, ecosystems, and marine life.

More acidic waters inhibit the formation of shells, skeletons, and coral reefs. Warmer waters harm coral reefs and alter the distribution, abundance, and productivity of many marine species. The rising temperature and changing chemistry of ocean water combine with other stresses, such as overfishing and coastal and marine pollution, to alter marine-based food production and harm fishing communities.

KEY MESSAGES: OCEANS

The rise in ocean temperature over the last century will persist into the future, with continued large impacts on climate, ocean circulation, chemistry, and ecosystems.

The ocean currently absorbs about a quarter of human-caused carbon dioxide emissions to the atmosphere, leading to ocean acidification that will alter marine ecosystems in dramatic yet uncertain ways.

Significant habitat loss will continue to occur due to climate change for many species and areas, including Arctic and coral reef ecosystems, while habitat in other areas and for other species will expand. These changes will consequently alter the distribution, abundance, and productivity of many marine species.

Rising sea surface temperatures have been linked with increasing levels and ranges of diseases in humans and marine life, including corals, abalones, oysters, fishes, and marine mammals.

Climate changes that result in conditions substantially different from recent history may significantly increase costs to businesses as well as disrupt public access and enjoyment of ocean areas.

In response to observed and projected climate impacts, some existing ocean policies, practices, and management efforts are incorporating climate change impacts. These initiatives can serve as models for other efforts and ultimately enable people and communities to adapt to changing ocean conditions.

As a nation, we depend on the oceans for seafood, recreation and tourism, cultural heritage, transportation of goods, and, increasingly, energy and other critical resources. The U.S. Exclusive Economic Zone extends 200 nautical miles seaward from the coasts, spanning an area about 1.7 times the land area of the continental United States. This vast region is host to a rich diversity of marine plants and animals and a wide range of ecosystems, from tropical coral reefs to Arctic waters covered with sea ice.

Oceans support vibrant economies and coastal communities with numerous businesses and jobs. More than 160 million people live in the coastal watershed counties of the U.S., and population in this zone is expected to grow in the future. The oceans help regulate climate, absorb carbon dioxide, and strongly influence weather patterns far into the continental interior. Ocean issues touch all of us in both direct and indirect ways.[1,2]

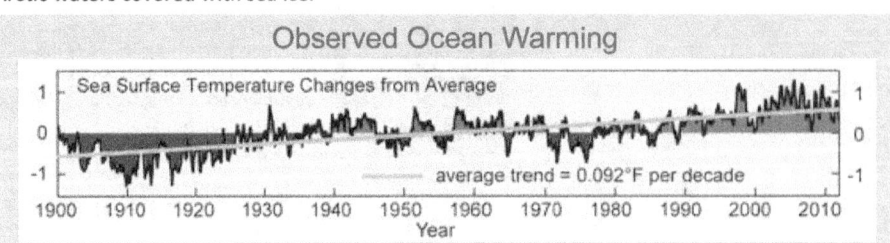

Observed Ocean Warming

Sea Surface Temperature Changes from Average

average trend = 0.092°F per decade

Sea surface temperatures for the ocean surrounding the U.S. and its territories have risen by more than 0.9°F over the past century. (Figure source: adapted from Chavez et al. 2011[3]).

APP

Changing climate conditions are already affecting these valuable marine ecosystems and the array of resources and services we derive from the sea. Some climate trends, such as rising seawater temperatures and ocean acidification, are common across much of the coastal areas and open ocean worldwide. The biological responses to climate change often vary from region to region, depending on the different combinations of species, habitats, and other attributes of local systems.

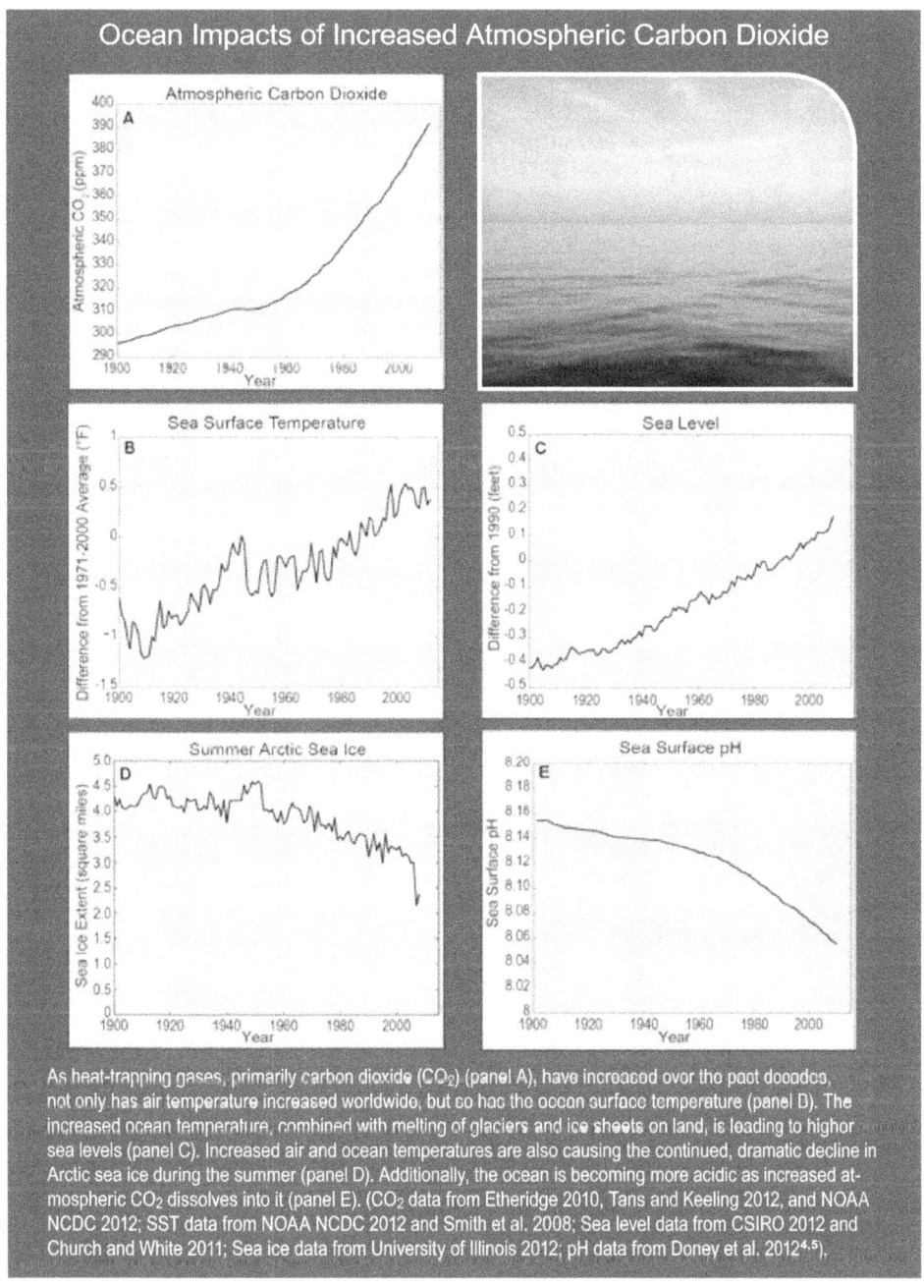

Ocean Impacts of Increased Atmospheric Carbon Dioxide

As heat-trapping gases, primarily carbon dioxide (CO_2) (panel A), have increased over the past decades, not only has air temperature increased worldwide, but so has the ocean surface temperature (panel B). The increased ocean temperature, combined with melting of glaciers and ice sheets on land, is leading to higher sea levels (panel C). Increased air and ocean temperatures are also causing the continued, dramatic decline in Arctic sea ice during the summer (panel D). Additionally, the ocean is becoming more acidic as increased atmospheric CO_2 dissolves into it (panel E). (CO_2 data from Etheridge 2010, Tans and Keeling 2012, and NOAA NCDC 2012; SST data from NOAA NCDC 2012 and Smith et al. 2008; Sea level data from CSIRO 2012 and Church and White 2011; Sea ice data from University of Illinois 2012; pH data from Doney et al. 2012[4,5]).

The oceans cover more than two-thirds of the Earth's surface and play a very important role in regulating the Earth's climate and in climate change. Today, the world's oceans absorb more than 90% of the heat trapped by increasing levels of CO_2 and other greenhouse gases in the atmosphere due to human activities. This extra energy warms the ocean, causing it to expand and sea levels to rise. Of the global sea level rise observed over the last 35 years, about 40% is due to this warming of the water. Most of the rest is due to the melting of glaciers and ice sheets. Ocean levels are projected to rise another 1 to 4 feet over this century, with the precise number largely depending on the amount of global temperature rise and polar ice sheet melt.

Observations from past climate combined with climate model projections of the future suggest that over the next 100 years the Atlantic Ocean's overturning circulation (known as the "Ocean Conveyor Belt") could slow down as a result of climate change. These ocean currents carry warm water northward across the equator in the Atlantic Ocean, warming the North Atlantic (and Europe) and cool-ing the South Atlantic. A slowdown of the Conveyor Belt would increase regional sea level rise along the east coast of the U.S. and change temperature patterns in Europe and rainfall in Africa and the Americas, but would not lead to global cooling.

Warming ocean waters also affect marine ecosystems like coral reefs, which can be very sensitive to temperature changes. When water temperatures become too high, coral expel the algae (called zooxanthellae) which help nourish them and give them their vibrant color. This is known as coral bleaching. If the high temperatures persist, the coral die.

Acidification

In addition to the warming, the acidity of seawater is increasing as a direct result of increasing atmospheric CO_2. Due to human-induced emissions, atmospheric CO_2 has risen by about 40% above pre-industrial levels.[5,6] About a quarter of this excess CO_2 has dissolved into the oceans, thereby changing seawater chemistry and decreasing pH (making seawater more acidic).[2,7] There has been about a 30% increase in surface ocean acidity since pre-industrial times.[8] Ocean acid-ification will continue in the future due to the interaction of atmospheric CO_2 and ocean water. Regional differences in ocean pH occur as a result of variability in regional or local conditions, such as upwelling that brings subsurface waters up to the surface.[9] Locally, coastal waters and estuaries can also exhibit acidification as the result of pollution and excess nutrient inputs.

More acidic waters create repercussions along the marine food chain. The chemi-cal changes caused by the uptake of CO_2 make it more difficult for living things to form and maintain calcium carbonate shells and skeletons and increases erosion of coral reefs,[10] resulting in alterations in marine ecosystems that will become more severe as present-day trends in acidifi-cation continue or accelerate.[11] Tropical corals are particularly susceptible to the combination of ocean acidification and ocean warming, which would threaten the rich and biologically diverse coral reef habitats. See page 33.

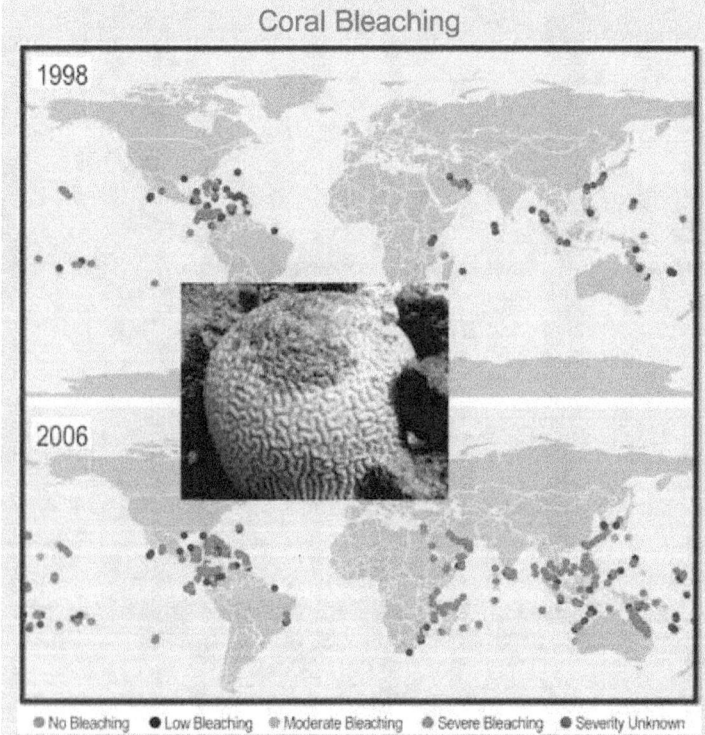

Coral Bleaching

1998

2006

● No Bleaching   ● Low Bleaching   ● Moderate Bleaching   ● Severe Bleaching   ● Severity Unknown

(Photo) Bleached brain coral; (Maps) The global extent and severity of mass coral bleaching have increased worldwide over the last decade. Red dots indicate severe bleaching. (Figure source: Marshall and Schuttenberg 2006;[12] Photo credit: NOAA).

Ocean Acidification Reduces Size of Clams

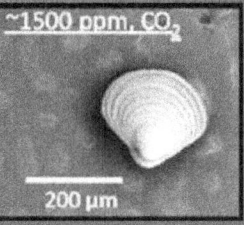

~250 ppm, CO$_2$ 200 µm

~390 ppm, CO$_2$ 200 µm

~750 ppm, CO$_2$ 200 µm

~1500 ppm, CO$_2$ 200 µm

These 36-day-old clams are a single species, *Mercenaria mercenaria*, grown in the laboratory under varying levels of carbon dioxide (CO$_2$) in the air. CO$_2$ is absorbed from the air by ocean water, acidifying the water and thus reducing the ability of juvenile clams to grow their shells. As seen in the photos, 36-day-old clams (measured in microns) grown under elevated CO$_2$ levels are smaller than those grown under lower CO$_2$ levels. The highest CO$_2$ level, about 1500 parts per million (ppm; far right), is higher than most projections for the end of this century but could occur locally in some estuaries. (Figure source: Talmage and Gobler 2010[13]).

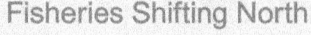

Fisheries Shifting North

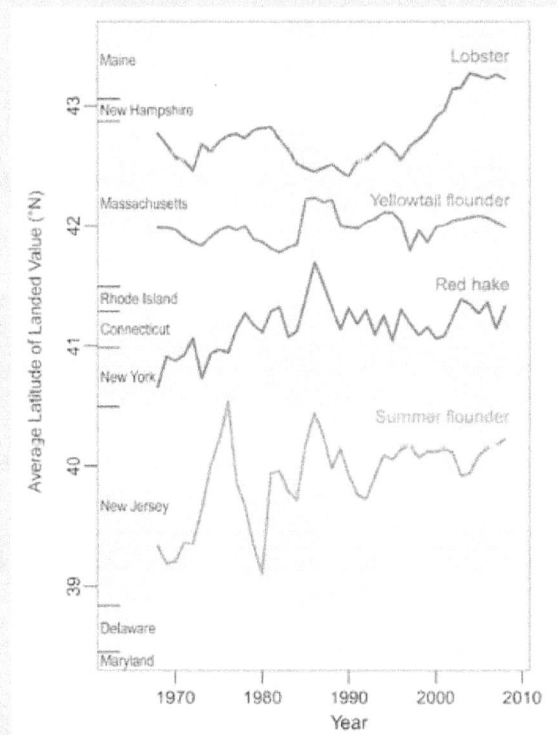

Ocean species are shifting northward along U.S. coastlines as ocean temperatures rise. As a result, over the past 40 years, more northern ports have gradually increased their landings of four marine species compared to earlier landings. While some species move northward out of an area, other species move in from the south. This kind of information can inform decisions about how to adapt to climate change. Such adaptations take time and have costs, as local knowledge and equipment are geared to the species that have long been present in an area. (Figure source: adapted from Pinsky and Fogerty 2012[19]).

Diseases

There has been a significant increase in reported incidences of disease in corals, urchins, mollusks, marine mammals, turtles, and echinoderms (a group of some 70,000 marine species including sea stars, sea urchins, and sand dollars) over the last several decades.[14,15] Increasing disease outbreaks in the ocean affecting ecologically important species, which provide critically important habitat for other species such as corals, algae, and eelgrass, have been linked with rising temperatures.[15,16,17] Disease increases mortality and can reduce abundance for affected populations as well as fundamentally change ecosystems by altering habitat or species relationships. For example, loss of eelgrass beds due to disease can reduce critical nursery habitat for several species of commercially important fish.[17,18]

12 RESPONSES

Planning for adaptation (to address and prepare for impacts) and mitigation (to reduce future climate change, for example by cutting emissions) is becoming more widespread, but current implementation efforts are insufficient to avoid increasingly negative social, environmental, and economic consequences.

Actions to reduce emissions, increase carbon uptake, adapt to a changing climate, and increase resilience to impacts that are unavoidable can improve public health, economic development, ecosystem protection, and quality of life.

Over the past few years, the focus moved from *"Is climate changing?"* to *"Can society manage unavoidable changes and avoid unmanageable changes?"*[1,2] Research demonstrates that both mitigation (efforts to reduce future climate changes) and adaptation (efforts to reduce the vulnerability of society to climate change impacts) are needed in order to minimize the damages from human-caused climate change and to adapt to the pace and ultimate magnitude of changes that will occur.[3] Adaptation and mitigation are closely linked; adaptation efforts will be more difficult, more costly, and less likely to succeed if significant mitigation actions are not taken.[2,4]

KEY MESSAGES: ADAPTATION

Substantial adaptation planning is occurring in the public and private sectors and at all levels of government; however, few measures have been implemented and those that have appear to be incremental changes.

Barriers to implementation of adaptation include limited funding, policy and legal impediments, and difficulty in anticipating climate related changes at local scales.

There is no "one-size fits all" adaptation, but there are similarities in approaches across regions and sectors. Sharing best practices, learning by doing, and iterative and collaborative processes including stakeholder involvement, can help support progress.

Climate change adaptation actions often fulfill other societal goals, such as sustainable development, disaster risk reduction, or improvements in quality of life, and can therefore be incorporated into existing decision-making processes.

Vulnerability to climate change is exacerbated by other stresses such as pollution and habitat fragmentation. Adaptation to multiple stresses requires assessment of the composite threats as well as tradeoffs amongst costs, benefits, and risks of available options.

The effectiveness of climate change adaptation has seldom been evaluated, because actions have only recently been initiated, and comprehensive evaluation metrics do not yet exist.

Adaptation actions can be implemented reactively, after changes in climate occur, or proactively, to prepare for a changing climate.[5] Proactively preparing can reduce the harm from certain climate change impacts, such as increasingly intense extreme events, shifting zones for agricultural crops, and rising sea levels, while also facilitating a more rapid and efficient response to changes as they happen.

FEDERAL: A November 2013 Executive Order calls for, among other things, modernizing federal programs to support climate resilient investments, managing lands and waters for climate preparedness and resilience, creating a Council on Climate Preparedness and Resilience, and the creation of a State, Local, and Tribal Leaders Task Force on Climate Preparedness and Resilience.[6] Federal agen-

cies are all required to plan for adaptation. Actions include coordinated efforts at the White House, regional and cross-sector efforts, agency-specific adaptation plans, and support for local-level adaptation planning and action.

STATE: States have become important actors in national climate change related efforts. State governments can create policies and programs that encourage or discourage adaptation at other governance scales (such as counties or regions)[7] through regulation and by serving as laboratories for innovation.[8] Although many of these actions are not specifically designed to address climate change, they often include climate adaptation components. Many state level climate change-specific adaptation actions focus on planning. As of winter 2012, at least 15 states had completed

climate adaptation plans; four states are in the process of writing their plans; and seven states have made recommendations to create state-wide adaptation plans.[9]

TRIBES: Tribal governments have been particularly active in assessing and preparing for the impacts of climate change. Some are using traditional knowledge gleaned from elders, stories, and songs and combining this knowledge with downscaled climate data to inform decision-making.[10] Others have integrated climate change into decision-making in major sectors, such as education, fisheries, social services, and human health.[11]

LOCAL: Most adaptation efforts to date have occurred at local and regional levels. A survey of 298 U.S. local governments shows 59% engaged in some form of adaptation planning.[12] Mechanisms used by local governments to prepare for climate change include: land-use planning; provisions to protect infrastructure and ecosystems; regulations related to the design and construction of buildings, road, and bridges; and preparation for emergency response and recovery.[13] Local adaptation planning and actions are unfolding in municipalities of different sizes. Regional agencies and regional aggregations of governments too are taking actions.[14]

BUSINESS: Many companies are concerned about how climate change will affect feedstock, water quality, infrastructure, core operations, supply chains, and customers' ability to use products and services.[15] Some companies are taking action to avoid risk and explore potential opportunities, such as: developing or expanding into new products, services, and operational areas; extending growing seasons and hours of operation; and responding to increased demand for existing products and services.[15,16]

NGOs: Non-governmental organizations have played significant roles in the national effort to prepare for climate change by providing assistance to stakeholders that includes planning guidance, implementation tools, explanations of climate information, best practices, and help with bridging the science-policy divide.

See regional sections of this Highlights report for additional examples of adaptation efforts. Selected federal, state, tribal, and local actions appear in the Adaptation chapter of the full National Climate Assessment.

Adaptation to climate change is in a nascent stage. The federal government is beginning to develop institutions and practices necessary to cope with climate change. While the federal government will remain the funder of emergency responses following extreme events for which communities were not adequately prepared, an emerging federal role is to enable and facilitate early adaptation within states, regions, local communities, and the public and private sectors.[5] The approaches include working to limit current institutional constraints to effective adaptation, funding pilot projects, providing useful and usable adaptation information – including disseminating best practices, and helping develop tools and techniques to evaluate successful adaptation.

Despite emerging efforts, the pace and extent of adaptation activities are not proportional to the risks to people, property, infrastructure, and ecosystems from climate change; important opportunities available during the normal course of planning and management of resources are also being overlooked. A number of state and local governments are engaging in adaptation planning, but most have not taken action to implement the plans.[17] Some companies in the private sector and numerous non-governmental organizations have also taken early action, particularly in capitalizing on the opportunities associated with facilitating adaptive actions. Actions and collaborations have occurred across all scales. At the same time, barriers to effective implementation continue to exist.

ADAPTATION EXAMPLE: The Southeast Florida Regional Compact

The Southeast Florida Regional Compact is a joint commitment among Broward, Miami-Dade, Palm Beach, and Monroe Counties to partner in reducing heat-trapping gas emissions and adapting to climate impacts, including in transportation, water resources, natural

Miami-Dade County staff leading workshop on incorporating climate change considerations in local planning.

resources, agriculture, and disaster risk reduction. Through the collaboration of county, state, and federal agencies, a comprehensive action plan was developed that includes hundreds of actions. Notable policies include regional collaboration to revise building codes and land development regulations to discourage new development or post-disaster redevelopment in vulnerable areas.[18]

KEY MESSAGES: MITIGATION

Carbon dioxide is removed from the atmosphere by natural processes at a rate that is roughly half of the current rate of emissions from human activities. Therefore, mitigation efforts that only stabilize global emissions will not reduce atmospheric concentrations of carbon dioxide, but will only limit their rate of increase. The same is true for other long-lived greenhouse gases.

To meet the lower emissions scenario (B1) used in this assessment, global mitigation actions would need to limit global carbon dioxide emissions to a peak of around 44 billion tons per year within the next 25 years and decline thereafter. In 2011, global emissions were around 34 billion tons, and have been rising by about 0.9 billion tons per year for the past decade. Therefore, the world is on a path to exceed 44 billion tons per year within a decade.

Over recent decades, the U.S. economy has emitted a decreasing amount of carbon dioxide per dollar of gross domestic product. Between 2008 and 2012, there was also a decline in the total amount of carbon dioxide emitted annually from energy use in the U.S. as a result of a variety of factors, including changes in the economy, the development of new energy production technologies, and various government policies.

Carbon storage in land ecosystems, especially forests, has offset around 17% of annual U.S. fossil fuel emissions of greenhouse gases over the past several decades, but this carbon "sink" may not be sustainable.

Both voluntary activities and a variety of policies and measures that lower emissions are currently in place at federal, state, and local levels in the U.S., even though there is no comprehensive national climate legislation. Over the remainder of this century, aggressive and sustained greenhouse gas emission reductions by the U.S. and by other nations will be needed to reduce global emissions to a level consistent with the lower scenario (B1) analyzed in this assessment.

The amount of future climate change will largely be determined by choices society makes about emissions. Lower emissions of heat trapping gases and particles mean less future warming and less severe impacts; higher emissions mean more warming and more severe impacts. Efforts to limit emissions or increase carbon uptake fall into a category of response options known as "mitigation."

Carbon dioxide accounted for 84% of total U.S. greenhouse gas emissions in 2011.[19] The vast majority (97%) of this CO_2 comes from energy use. Thus, the most direct way to reduce future climate change is to reduce emissions from the energy sector by using energy more efficiently and switching to lower carbon energy sources.

In 2011, 41% of U.S. carbon dioxide emissions were attributable to liquid fuels (petroleum), followed closely by solid fuels (principally coal in electric generation), and to a lesser extent by natural gas.[19] Electric power generation (coal and gas) and transportation (petroleum) are the sectors predominantly responsible.

Achieving the lower emissions path (B1) analyzed in this assessment would require substantial decarbonization of the global economy by the end of this century, implying a fundamental transformation of the global energy system. The principal types of national actions that could effect such changes include putting a price on emissions, setting regulations and standards for activities that cause

emissions, changing subsidy programs, and direct federal expenditures. Market-based approaches include cap-and-trade programs that establish markets for trading emissions permits, analogous to the Clean Air Act provisions for sulfur dioxide reductions.

None of these price-based measures has been implemented at the national level in the U.S., though cap-and-trade systems are in place in California and in the Northeast's Regional Greenhouse Gas Initiative. A wide range of governmental actions are underway at federal, state, regional, and city levels using other measures, as are voluntary efforts, that can reduce the U.S. contribution to total global emissions. Many, if not most of these programs are motivated by other policy objectives – energy, transportation, and air pollution – but some are directed specifically at greenhouse gas emissions, including:

- **Energy Efficiency:** Reduction in CO_2 emissions from energy end-use and infrastructure through the adoption of energy-efficient components and systems – including buildings, vehicles, manufacturing processes, applicances, and electric grid systems;

- **Low-Carbon Energy Sources:** Reduction of CO_2 emissions from energy supply through the promotion of renewables (such as wind, solar, and bioenergy), nuclear energy, and coal and natural gas electric generation with carbon capture and storage; and

Programs underway that reduce carbon dioxide emissions include the promotion of solar, nuclear, and wind power, and efficient vehicles.

- **Non-CO$_2$ Emissions:** Reduction of emissions of non-CO$_2$ greenhouse gases and black carbon (soot); for example, by lowering methane emissions from energy and waste, transitioning to climate-friendly alternatives to HFCs, cutting methane and nitrous oxide emissions from agriculture, and improving combustion efficiency and means of particulate capture.

Federal Actions

The Federal Government has implemented a number of measures that promote energy efficiency, clean technologies, and alternative fuels.[20] Sample federal measures are provided in Table 27.1 in the Mitigation chapter in the full report. These actions include greenhouse gas regulations, other rules and regulations with climate co-benefits, various standards and subsidies, research and development, and federal procurement practices.

For example, the Environmental Protection Agency has the authority to regulate greenhouse gas emissions under the Clean Air Act. The Department of Energy provides most of the funding for energy research and development, and also regulates the efficiency of appliances.

The Administration's Climate Action Plan[21] builds on these activities with a broad range of mitigation, adaptation, and preparedness measures. The mitigation elements of the plan are in part a response to the commitment made during the 2010 Cancun Conference of the Parties of the United Nations Framework Convention on Climate Change to reduce U.S. emissions of greenhouse gases by about 17% below 2005 levels by 2020. Actions proposed in the Plan include:

- limiting carbon emissions from both new and existing power plants;
- continuing to increase the stringency of fuel economy standards for automobiles and trucks;
- continuing to improve energy efficiency in the buildings sector;
- reducing the emissions of non-CO$_2$ greenhouse gases through a variety of measures;
- increasing federal investments in cleaner, more efficient energy sources for both power and transportation; and
- identifying new approaches to protect and restore our forests and other critical landscapes, in the presence of a changing climate.

CO-BENEFITS FOR AIR POLLUTION AND HUMAN HEALTH

Actions to reduce greenhouse gas emissions can yield co-benefits for objectives apart from climate change, such as energy security, ecosystem services, and biodiversity.[22] In particular, there are health co-benefits from reductions in air pollution. Because greenhouse gases and other air pollutants share common sources, particularly from fossil fuel combustion, actions to reduce greenhouse gas emissions also reduce other air pollutants.

The human health benefits can be immediate and local, in contrast to the long-term and widespread effects of climate change.[23] These efforts have been found to be cost effective.[23,24] Methane reductions have also been shown to generate health benefits from reduced ground-level ozone.[25]

Actions to reduce greenhouse gases can also reduce other air pollutants, yielding human health benefits.

City, State, and Regional Actions

Jurisdiction for greenhouse gases and energy policies is shared between the Federal government and states.[26] For example, states regulate the distribution of electricity and natural gas to consumers, while the Federal Energy Regulatory Commission regulates wholesale sales and transportation of natural gas and electricity. Many states have adopted climate initiatives as well as energy policies that reduce greenhouse gas emissions. For a survey of many of these state activities, see Table 27.2 in the full report. Many cities are taking similar actions.

The most ambitious state activity is California's Global Warming Solutions Act, with a goal of reducing greenhouse gas emissions to 1990 levels by 2020. The program caps emissions and uses a market-based system of trading in emissions credits, as well as a number of regulatory actions. The most well-known, multi-state effort has been the Regional Greenhouse Gas Initiative (RGGI), formed by 10 northeastern and Mid-Atlantic states (though New Jersey exited in 2011). RGGI is a cap-and-trade system in the power sector directing revenue from allowance auctions to investments in efficiency and renewable energy.

Voluntary Actions

Corporations, individuals, and non-profit organizations have initiated a host of voluntary actions, including:

- The Carbon Disclosure Project enables companies to measure, disclose, manage, and share climate change and water-use information. Some 650 U.S. signatories include banks, pension funds, asset managers, insurance companies, and foundations.
- More than 1,055 municipalities from all 50 states have signed the U.S. Mayors Climate Protection Agreement,[27] and many of these communities are actively implementing strategies to reduce their emissions.
- Federal voluntary programs include Energy STAR, a labeling program that, among other things, identifies energy efficient products for use in residences and commercial and industrial buildings.

Managing Land for Mitigation

Mitigation can involve increasing the uptake of carbon through various means of expanding carbon sinks on land through management of forests and soils.

SELECTED MITIGATION MEASURES

Existing federal laws and regulations to reduce emissions include:

Emissions Standards for Vehicles and Engines

- For light-duty vehicles, rules establishing standards for 2012-2016 model years and 2017-2025 model years.
- For heavy- and medium-duty trucks, a rule establishing standards for 2014-2018 model years.

Appliance and Building Efficiency Standards

- Energy efficiency standards and test procedures for residential, commercial, industrial, lighting, and plumbing products.
- Model residential and commercial building energy codes, and technical assistance to state and local governments, and non-governmental organizations.

Financial Incentives for Efficiency and Alternative Fuels and Technology

- Weatherization assistance for low-income households, tax incentives for commercial and residential buildings and efficient appliances, and support for state and local efficiency programs.

Weatherization can include installing more efficient windows to save energy.

KEY MESSAGES: DECISION SUPPORT

Decisions about how to address climate change can be complex, and responses will require a combination of adaptation and mitigation actions. Decision-makers – whether individuals, public officials, or others – may need help integrating scientific information into adaptation and mitigation decisions.

To be effective, decision support processes need to take account of the values and goals of the key stakeholders, evolving scientific information, and the perceptions of risk.

Many decision support processes and tools are available. They can enable decision-makers to identify and assess response options, apply complex and uncertain information, clarify trade-offs, strengthen transparency, and generate information on the costs and benefits of different choices.

Ongoing assessment processes should incorporate evaluation of decision support tools, their accessibility to decision-makers, and their application in decision processes in different sectors and regions.

Steps to improve collaborative decision processes include developing new decision support tools and building human capacity to bridge science and decision-making.

As a result of human-induced climate change, historically successful strategies for managing climate-sensitive resources and infrastructure will become less effective over time. Decision support processes and tools can help structure decision-making, organize and analyze information, and build consensus around options for action.

Although decision-makers routinely make complex decisions under uncertain conditions, decision-making in the context of climate change can be especially challenging. Reasons include the rapid pace of changes, long time lags between human activities and response of the climate system, the high economic and political stakes, the number and diversity of potentially affected stakeholders, the need to incorporate uncertain scientific information of varying confidence levels, and the values of stakeholders and decision-makers.[28,29] The social, economic, psychological, and political dimensions of these decisions underscore the need for ways to improve communication of scientific information and uncertainties and to help decision-makers assess risks and opportunities.

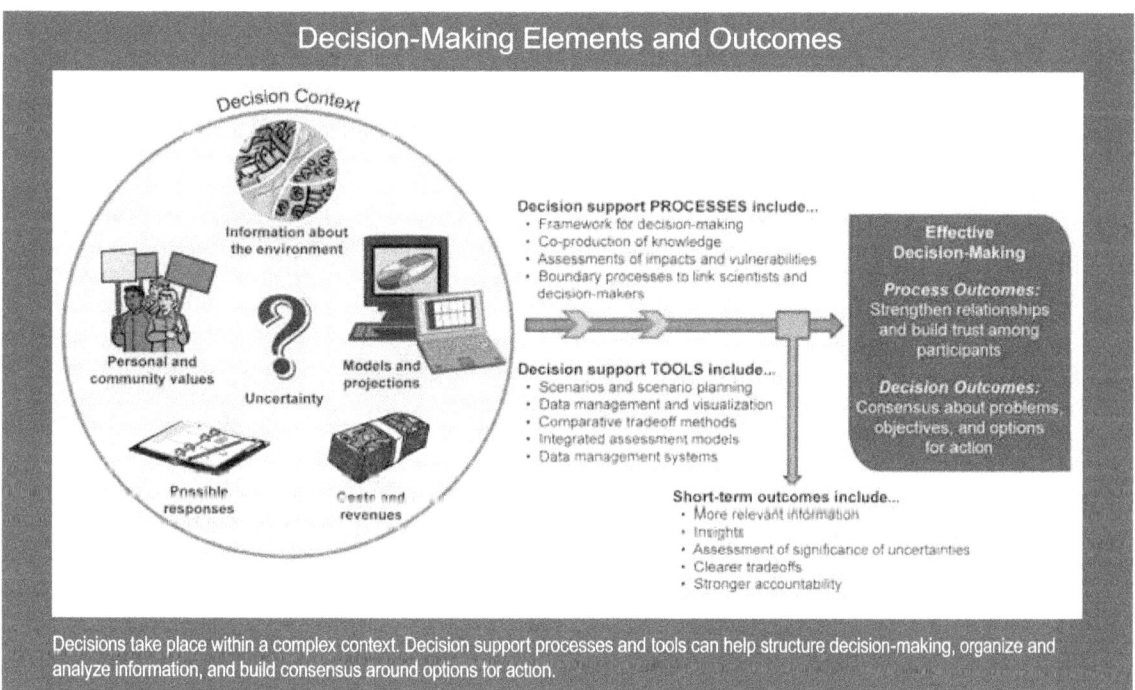

Decisions take place within a complex context. Decision support processes and tools can help structure decision-making, organize and analyze information, and build consensus around options for action.

Finding 12: RESPONSES

Collaboration: The importance of both scientific information and societal considerations suggests the need for the public, technical experts, and decision-makers to engage in mutual shared learning and shared production of relevant knowledge.[29,30]

Uncertainty: An "iterative adaptive risk management framework" is useful for decisions about adaptation and ways to reduce future climate change, especially given uncertainties and ongoing advances in scientific understanding.[31] An idealized iterative adaptive risk management process includes clearly defining the issue, establishing decision criteria, identifying and incorporating relevant information, evaluating options, and monitoring and revisiting effectiveness.

Decision-Making Framework

Define the issue

Establish decision criteria

Access information, assess risk and available decision support

Enhance understanding

Integrate, evaluate, assess tradeoffs, and decide

Implement

Monitor

Analyze and re-evaluate decision

Collaboration & Learning (Decision-makers, scientists, and stakeholders)

This illustration highlights several stages of a well-structured decision-making process. (Figure source: adapted from NRC 2010 and Willows and Connell 2003[31]).

Risk Management: Making effective climate-related decisions requires balance among actions intended to manage, reduce, and transfer risk. Risks are threats to life, health and safety, the environment, economic well-being, and other things of value. Methods such as multiple criteria analysis, valuation of both risks and opportunities, and scenarios can help to combine experts' assessment of climate change risks with public perception of these risks.[32]

Decision Support Case Study: Denver Water

Climate change is one of the biggest challenges facing the Denver Water system. Due to recent and anticipated effects of climate variability and change on water availability, Denver Water faces the challenge of weighing alternative response strategies and is looking at developing options to help meet more challenging future conditions.

Denver Water is using scenario planning in its long-range planning process (looking out to 2050) to consider a range of plausible futures involving climate change, demographic and water use changes, and economic and regulatory changes. The strategy focuses on keeping as many future options open as possible while trying to ensure reliability of current supplies.

The next step for Denver Water is to explore a more technical approach to test their existing plan and identified options against multiple climate change scenarios. Following a modified robust decision-making approach,[33] Denver Water will test and hedge its plan and options until those options demonstrate that they can sufficiently handle a range of projected climate conditions.

REGIONS

Evidence of climate change can be found in every region, and impacts are visible in every state.

Americans are seeing changes such as species moving northward, increases in invasive species and insect outbreaks, and changes in the length of the growing season. In many cities, impacts to the urban environment are closely linked to the changing climate, with increased flooding, greater incidence of heat waves, and diminished air quality. Along most of our coastlines, increasing sea levels and associated threats to coastal areas and infrastructure are becoming a common experience.

The pages that follow provide a summary of changes and impacts that are observed and anticipated in each of the eight regions of the United States, as well as in rural and coastal areas.

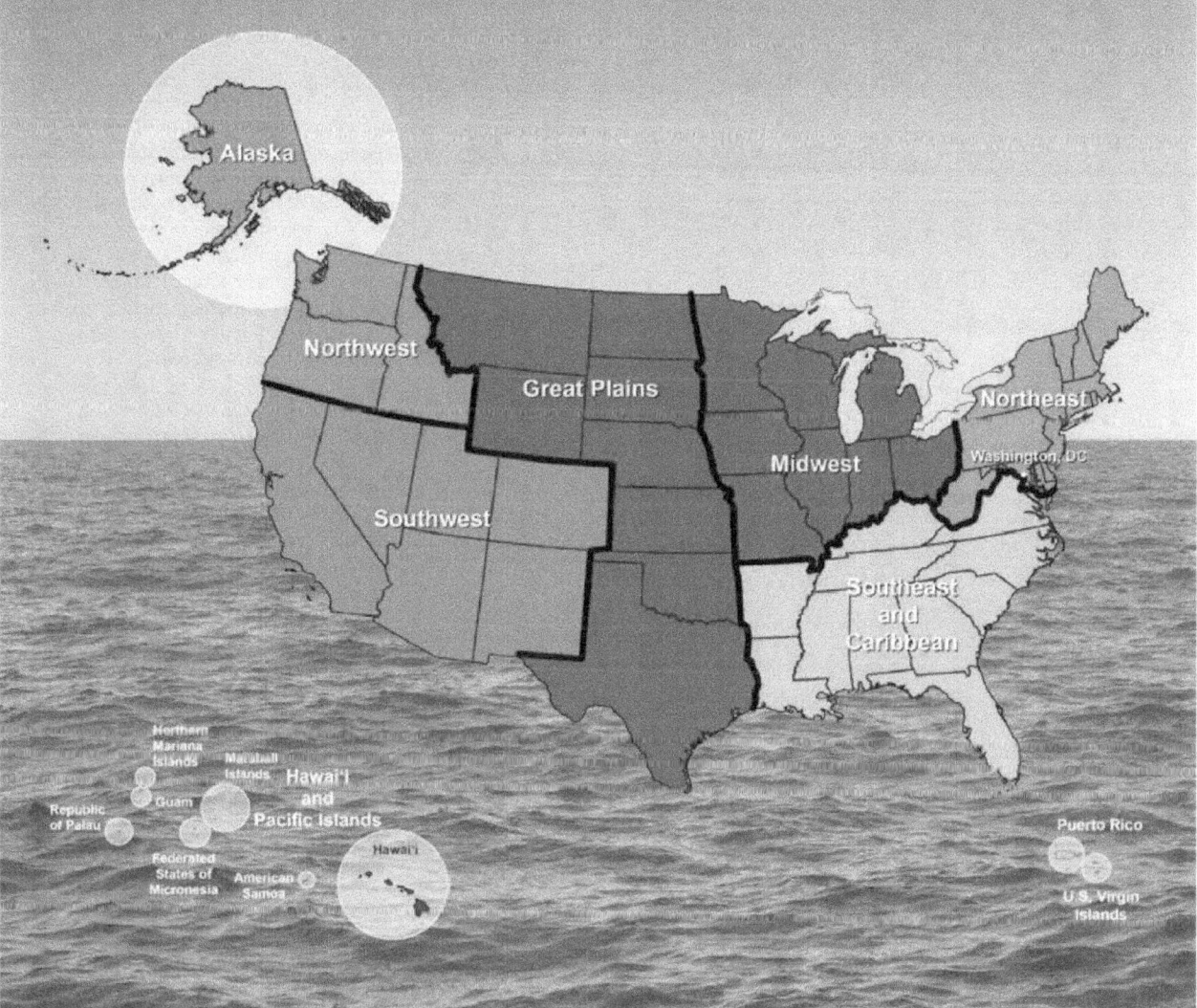

NORTHEAST

KEY MESSAGES

Heat waves, coastal flooding, and river flooding will pose a growing challenge to the region's environmental, social, and economic systems. This will increase the vulnerability of the region's residents, especially its most disadvantaged populations.

Infrastructure will be increasingly compromised by climate-related hazards, including sea level rise, coastal flooding, and intense precipitation events.

Agriculture, fisheries, and ecosystems will be increasingly compromised over the next century by climate change impacts. Farmers can explore new crop options, but these adaptations are not cost- or risk-free. Moreover, adaptive capacity, which varies throughout the region, could be overwhelmed by a changing climate.

While a majority of states and a rapidly growing number of municipalities have begun to incorporate the risk of climate change into their planning activities, implementation of adaptation measures is still at early stages.

Urban Heat Island

Temperature (°F)

64.4 66.9 69.9 72.1 74.8 77.4 79.9 82.6 85.1 87.6 90.3 92.8 95.4 98.1 100.6 103.3 105.8

Surface temperatures in New York City on a summer's day show the "urban heat island," with temperatures in populous urban areas being approximately 10°F higher than the forested parts of Central Park. Dark blue reflects the colder waters of the Hudson and East Rivers. (Figure source: Center for Climate Systems Research, Columbia University).

Sixty-four million people are concentrated in the Northeast. The high-density urban coastal corridor from Washington, D.C., north to Boston is one of the most developed environments in the world. It contains a massive, complex, and long-standing network of supporting infrastructure. The Northeast also has a vital rural component, including large expanses of sparsely populated but ecologically and agriculturally important areas.

Although urban and rural regions in the Northeast are profoundly different, they both include populations that are highly vulnerable to climate hazards and other stresses. The region depends on aging infrastructure that has already been stressed by climate hazards including heat waves and heavy downpours. The Northeast has experienced a greater recent increase in extreme precipitation than any other region in the U.S.; between 1958 and 2010, the Northeast saw more than a 70% increase in the amount of precipitation falling in very heavy events (defined as the heaviest 1% of all daily events).[1] This increase, combined with coastal and riverine flooding due to sea level rise and storm surge, creates increased risks. For all of these reasons, public health, agriculture, transportation, communications, and energy systems in the Northeast all face climate-related challenges.

Hurricane Vulnerability

Hurricanes Irene and Sandy demonstrated the region's vulnerability to extreme weather events and the potential for adaptation to reduce impacts. Hurricane Irene produced a broad swath of very heavy rain (greater than 5 inches in total and 2 to 3 inches per hour in some locations) from southern Maryland to northern Vermont from August 27 to 29, 2011. These heavy rains were part of a broader pattern of wet weather preceding the storm that exacerbated the flooding.

In anticipation of Irene, the New York City mass transit system was shut down, and 2.3 million coastal residents in Delaware, New Jersey, and New York faced mandatory evacuations. But inland impacts, especially in upstate New York and in central and southern Vermont, were most severe. Flash flooding washed out roads and bridges, undermined railroads, brought down trees and power lines, flooded homes and businesses, and damaged floodplain forests. Hazardous wastes were released in a number of areas, and 17 municipal wastewater treatment plants were breached by the floodwaters. Crops were flooded and many towns and villages were isolated for days.

Hurricane Sandy, which hit the East Coast in October 2012, caused massive coastal damage from storm surge and flooding. Sandy was responsible for approximately 150 deaths, about half of those in the Northeast, and monetary impacts on coastal areas, especially in New Jersey, New York, and Connecticut estimated at $60 to $80 billion.[2,3] Floodwaters inundated subway tunnels in New York City, 8.5 million people were without power, and an estimated 650,000 homes were damaged or destroyed.[7]

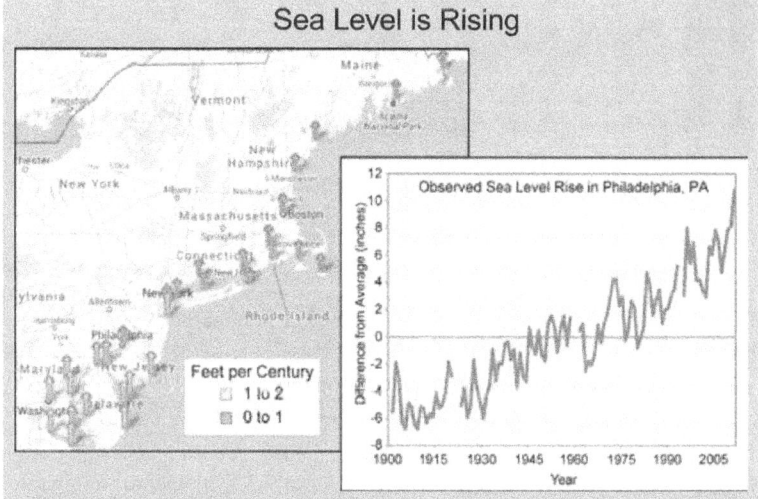

Sea Level is Rising

Observed Sea Level Rise in Philadelphia, PA

Feet per Century
- 1 to 2
- 0 to 1

Rising sea levels are already affecting coastal cities in the Northeast, and projections suggest that impacts will be widespread. The map on the left shows local sea level trends in the Northeast region. The length of the arrows varies with the length of the time series for each tide gauge location. (Figure source: NOAA). The graph at the right shows observed sea level rise in Philadelphia, which has increased by 1.2 feet over the past century, significantly exceeding the global average of 8 inches, increasing the risk of impacts to critical urban infrastructure in low-lying areas. (Data from Permanent Service for Mean Sea Level[5]).

SELECTED ADAPTATION EFFORTS

This one-acre stormwater wetland was constructed in Philadelphia to treat stormwater runoff in an effort to improve drinking water quality while minimizing the impacts of storm-related flows on natural ecosystems.

The City of Philadelphia is greening its combined sewer infrastructure to protect rivers, reduce greenhouse gas emissions, improve air quality, and enhance adaptation to a changing climate.[4]

Officials in coastal Maine are working with the statewide Sustainability Solutions Initiative to identify how culverts that carry stormwater can be maintained and improved, in order to increase resiliency to more frequent extreme precipitation events. This includes actions such as using larger culverts to carry water from major storms.[5]

SOUTHEAST AND CARIBBEAN

KEY MESSAGES

Sea level rise poses widespread and continuing threats to both natural and built environments and to the regional economy.

Increasing temperatures and the associated increase in frequency, intensity, and duration of extreme heat events will affect public health, natural and built environments, energy, agriculture, and forestry.

Decreased water availability, exacerbated by population growth and land-use change, will continue to increase competition for water and affect the region's economy and unique ecosystems.

The Southeast and Caribbean region is exceptionally vulnerable to sea level rise, extreme heat events, hurricanes, and decreased water availability. The geographic distribution of these impacts and vulnerabilities is uneven, since the region encompasses a wide range of environments, from the Appalachian Mountains to the coastal plains. The region is home to more than 80 million people and some of the fastest-growing metropolitan areas,[1] three of which are along the coast and vulnerable to sea level rise and storm surge. The Gulf and Atlantic coasts are major producers of seafood and home to seven major ports[2] that are also vulnerable. The Southeast is a major energy producer of coal, crude oil, and natural gas, and is the highest energy user of any of the National Climate Assessment regions.[2]

The Southeast warmed during the early part of last century, cooled for a few decades, and is now warming again. Temperatures across the region are expected to increase in the future. Major consequences include significant increases in the number of hot days (95°F or above) and decreases in freezing events. Higher temperatures contribute to the formation of harmful air pollutants and allergens.[3] Higher temperatures are also projected to reduce livestock and crop productivity.[4] Climate change is expected to increase harmful blooms of algae and several disease-causing agents in inland and coastal waters.[5] The number of Category 4 and 5 hurricanes in the North Atlantic and the amount of rain falling in very heavy precipitation events have increased over recent decades, and further increases are projected.

Southeast Temperature: Observed and Projected

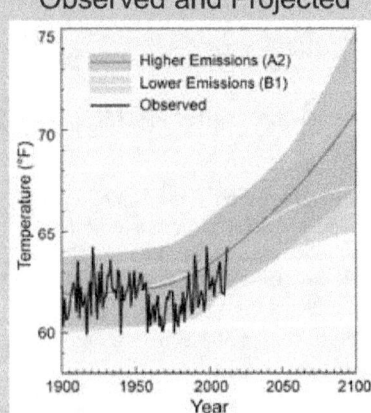

Temperature projections compared to observed temperatures from 1901-1960 for two emissions scenarios, one assuming substantial emissions reductions (B1) and the other continued growth in emissions (A2). For each scenario, shading shows range of projections and line shows a central estimate. (Figure source: adapted from Kunkel et al. 2013[6]).

Billion Dollar Weather/Climate Disasters 1980-2012

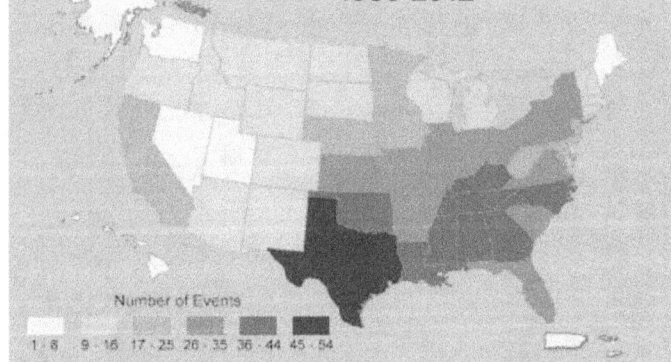

This map summarizes the number of times over the past 30 years that each state has been affected by weather and climate events that have resulted in more than a billion dollars in damages. The Southeast has been affected by more billion-dollar disasters than any other region. The primary disaster type for coastal states such as Florida is hurricanes, while interior and northern states in the region also experience sizeable numbers of tornadoes and winter storms. (Figure source: NOAA NCDC[7]).

Global sea level rose about eight inches in the last century and is projected to rise another 1 to 4 feet in this century. Large numbers of southeastern cities, roads, railways, ports, airports, oil and gas facilities, and water supplies are vulnerable to the impacts of sea level rise. Major cities like New Orleans, with roughly half of its population below sea level,[8] Miami, Tampa, Charleston, and Virginia Beach are among those most at risk.[9]

As a result of current sea level rise, the coastline of Puerto Rico around Rincòn is being eroded at a rate of 3.3 feet per year.[10] Puerto Rico has one of the highest population densities in the world, with 56% of the population living in coastal municipalities.[10]

Sea level rise and storm surge can have impacts far beyond the area directly affected. Sea level rise combines with other climate-related impacts and existing pressures such as land subsidence, causing significant economic and ecological implications. According to a recent study co-sponsored by a regional utility, coastal areas in Alabama, Mississippi, Louisiana, and Texas already face losses that annually average $14 billion from hurricane winds, land subsidence, and sea level rise. Losses for the 2030 timeframe could reach $23 billion assuming a nearly 3% increase in hurricane wind speed and just under 6 inches of sea level rise. About 50% of the increase in losses is related to climate change.[11]

Louisiana State Highway 1, heavily used for delivering critical oil and gas resources from Port Fourchon, is sinking, at the same time sea level is rising, resulting in more frequent and more severe flooding during high tides and storms.[12] A 90-day shutdown of this road would cost the nation an estimated $7.8 billion.[13]

Freshwater supplies from rivers, streams, and groundwater sources near the coast are at risk from accelerated saltwater intrusion due to higher sea levels. Porous aquifers in some areas make them particularly vulnerable to saltwater intrusion.[14] For example, officials in the city of Hallandale Beach, Florida, have already abandoned six of their eight drinking water wells.[15]

Continued urban development and increases in irrigated agriculture will increase water demand while higher temperatures will increase evaporative losses. All of these factors will combine to reduce the availability of water in the Southeast. Severe water stress is projected for many small Caribbean islands.[16]

Vulnerability to Sea Level Rise

Virginia Beach

Charleston

Tampa

New Orleans

Miami

Low   Moderate   High   Very High

The map shows the relative risk as sea level rises using a Coastal Vulnerability Index calculated based on tidal range, wave height, coastal slope, shoreline change, landform and processes, and historical rate of relative sea level rise. The approach combines a coastal system's susceptibility to change with its natural ability to adapt to changing environmental conditions, and yields a relative measure of the system's natural vulnerability to the effects of sea level rise. (Data from Hammar-Klose and Thieler 2001[17]).

SELECTED ADAPTATION EFFORTS

Clayton County, Georgia's innovative water recycling project enabled it to maintain abundant water supplies, with reservoirs at or near capacity, during the 2007-2008 drought, while neighboring Lake Lanier, the water supply for Atlanta, was at record lows. The project involved a series of constructed wetlands (see photo) used as the final stage of a wastewater treatment process that recharges groundwater and supplies surface reservoirs. The county has also implemented water efficiency and leak detection programs.[18]

In other adaptation efforts, the North Carolina Department of Transportation is raising U.S. Highway 64 across the Albemarle-Pamlico Peninsula by four feet, which includes 18 inches to allow for higher future sea levels.[19]

For another example, see page 63 for a description of the Southeast Florida Regional Compact's plans to reduce heat-trapping gas emissions and adapt to climate change impacts.

MIDWEST

KEY MESSAGES

In the next few decades, longer growing seasons and rising carbon dioxide levels will increase yields of some crops, though those benefits will be progressively offset by extreme weather events. Though adaptation options can reduce some of the detrimental effects, in the long term, the combined stresses associated with climate change are expected to decrease agricultural productivity.

The composition of the region's forests is expected to change as rising temperatures drive habitats for many tree species northward. The role of the region's forests as a net absorber of carbon is at risk from disruptions to forest ecosystems, in part due to climate change.

Increased heat wave intensity and frequency, increased humidity, degraded air quality, and reduced water quality will increase public health risks.

The Midwest has a highly energy-intensive economy with per capita emissions of greenhouse gases more than 20% higher than the national average. The region also has a large and increasingly utilized potential to reduce emissions that cause climate change.

Extreme rainfall events and flooding have increased during the last century, and these trends are expected to continue, causing erosion, declining water quality, and negative impacts on transportation, agriculture, human health, and infrastructure.

Climate change will exacerbate a range of risks to the Great Lakes, including changes in the range and distribution of certain fish species, increased invasive species and harmful blooms of algae, and declining beach health. Ice cover declines will lengthen the commercial navigation season.

The Midwest's agricultural lands, forests, Great Lakes, industrial activities, and cities are all vulnerable to climate variability and climate change. Climate change will tend to amplify existing risks climate poses to people, ecosystems, and infrastructure. Direct effects will include increased heat stress, flooding, drought, and late spring freezes. Climate change also alters pests and disease prevalence, competition from non-native or opportunistic native species, ecosystem disturbances, land-use change, landscape fragmentation, atmospheric and watershed pollutants, and economic shocks such as crop failures, reduced yields, or toxic blooms of algae due to extreme weather events. These added stresses, together with the

Projected Climate Change

Change in Days Above 95°F

Difference in Number of Days

0   5   10   15   20   25

Temperatures above 95°F are associated with negative human health impacts and suppressed agricultural yields. The frequency of these days is projected to increase by mid-century.

Change in Cooling Degree Days

Difference in Number of Cooling Degree Days

0   75   150   225   300   375

Cooling degree days (a measure of energy demand for air conditioning) are also projected to increase, leading to potential increases in the seasonality and annual total electricity demand.

Change in Heavy Precipitation

Difference in Number of Days

0   0.3   0.6   0.8   1.2   1.5

The frequency of days with very heavy precipitation (the wettest 2% of days) is also projected to increase, raising the risk of floods and nutrient pollution.

Projections above from global climate models are shown for 2041-2070 as compared to 1971-2000 under an emissions scenario that assumes continued increases in heat-trapping gases (A2 scenario). (Figure source: NOAA NCDC / CICS-NC)

direct effects of climate change, are projected to alter ecosystem and socioeconomic patterns and processes in ways that most people in the region would consider detrimental.

Most of the Midwest's population lives in urban environments. Climate change may intensify other stresses on urban dwellers and vegetation, including increased atmospheric pollution, heat island effects, a highly variable water cycle, and frequent exposure to new pests and diseases. Further, many of the cities have aging infrastructure and are particularly vulnerable to climate change related flooding and life-threatening heat waves. The increase in heavy downpours has contributed to the discharge of untreated sewage due to excess water in combined sewage-overflow systems in a number of cities in the Midwest.[1]

Much of the region's fisheries, recreation, tourism, and commerce depend on the Great Lakes and expansive northern forests, which already face pollution and invasive species pressures – pressures exacerbated by climate change.

Extreme weather events will influence future crop yields more than changes in average temperature or annual precipitation. High temperatures during early spring, for example, can decimate fruit crop production[2] when early heat causes premature plant budding that exposes flowers to later cold injury, as happened in 2002, and again in 2012, to Michigan's $60 million tart cherry crop. Springtime cold air outbreaks are projected to continue to occur throughout this century.[3]

Any increased productivity of some crops due to higher temperatures, longer growing seasons, and elevated carbon dioxide concentrations could be offset by water limitations and other stressors.[4] Heat waves during pollination of field crops such as corn and soybean also reduce yields.[5] Wetter springs may reduce crop yields and profits,[6] especially if growers are forced to switch to late-planted, shorter-season varieties.

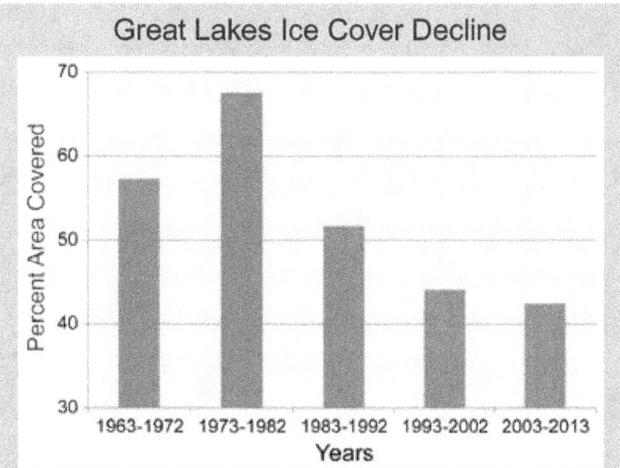

Great Lakes Ice Cover Decline

Great Lakes ice coverage has declined substantially, as shown by these decade averages of annual maximum ice coverage since reliable measurements began, although there is substantial variability from year to year. Less ice, coupled with more frequent and intense storms,[7] leaves shores vulnerable to erosion and flooding and could harm property and fish habitat.[8] Reduced ice cover also has the potential to lengthen the shipping season.[9] The navigation season increased by an average of eight days between 1994 and 2011. Increased shipping days benefit commerce but could also increase shoreline scouring and bring in more invasive species.[9,10] (Figure source: Data updated from Bai and Wang 2012[11]).

SELECTED ADAPTATION EFFORTS

The city of Cedar Falls' new floodplain ordinance expands zoning restrictions from the 100-year floodplain to the 500-year floodplain to better reflect the flood risks experienced by this and other Midwest cities during the 2008 floods.[12]

Cedar Rapids has also taken significant steps to reduce future flood damage, with buyouts of more than 1,000 properties, and numerous buildings adapted with flood protection measures.

Some cities have begun to incorporate adaptation planning for a range of climate change impacts. Chicago was one of the first cities to officially integrate climate adaptation into a citywide plan. Since the Climate Adaptation Plan's release, a number of strategies have been implemented to help the city manage heat, protect forests, and enhance green design, using techniques such as green roofs.[13]

GREAT PLAINS

KEY MESSAGES

Rising temperatures are leading to increased demand for water and energy. In parts of the region, this will constrain development, stress natural resources, and increase competition for water among communities, agriculture, energy production, and ecological needs.

Changes to crop growth cycles due to warming winters and alterations in the timing and magnitude of rainfall events have already been observed; as these trends continue, they will require new agriculture and livestock management practices.

Landscape fragmentation is increasing, for example, in the context of energy development activities in the northern Great Plains. A highly fragmented landscape will hinder adaptation of species when climate change alters habitat composition and timing of plant development cycles.

Communities that are already the most vulnerable to weather and climate extremes will be stressed even further by more frequent extreme events occurring within an already highly variable climate system.

The magnitude of expected changes will exceed those experienced in the last century. Existing adaptation and planning efforts are inadequate to respond to these projected impacts.

The Great Plains is a diverse region where climate is woven into the fabric of life. Daily, monthly, and yearly variations in the weather can be dramatic and challenging. The region experiences multiple climate and weather hazards, including floods, droughts, severe storms, tornadoes, hurricanes, and winter storms. In much of the Great Plains, too little precipitation falls to replace that needed by humans, plants, and animals. These variable conditions already stress communities and cause billions of dollars in damage. Climate change will add to both stress and costs.

The people of the Great Plains historically have adapted to this challenging climate. Although projections suggest more frequent and more intense droughts, heavy downpours, and heat waves, people can reduce vulnerabilities through the use of new technologies, community-driven policies, and the judicious use of resources. Efforts to reduce greenhouse gas emissions and adapt to climate change can be locally driven, cost effective, and beneficial for local economies and ecosystem services.

Even small shifts in timing of plant growth cycles caused by climate change can disrupt ecosystem functions like predator-prey relationships or food availability. While historic bison herds migrated to adapt to changing conditions, habitats are now fragmented by roads, agriculture, and structures, inhibiting similar large-scale migration.[1]

Increases in heavy downpours contribute to flooding.

The trend toward more dry days and higher temperatures across the Southern Plains will increase evaporation, decrease water supplies, reduce electricity transmission capacity, and increase cooling demands. These changes will add stress to limited water resources and affect management choices related to irrigation, municipal use, and energy generation.[2] Increased drought frequency and intensity can turn marginal lands into deserts.

Changing extremes in precipitation are projected across all seasons, including higher likelihoods of both increasing heavy rain and snow events[3] and more intense droughts.[4] Winter and spring precipitation and heavy downpours are both projected to increase in the

A Texas State Park police officer walks across a cracked lakebed in August 2011. This lake once spanned more than 5,400 acres.

north, leading to increased runoff and flooding that will reduce water quality and erode soils. Increased snowfall, rapid spring warming, and intense rainfall can combine to produce devastating floods, as is already common along the Red River of the North. More intense rains will also contribute to urban flooding.

Expectations of more precipitation in the northern Great Plains and less in the southern Great Plains were strongly manifest in 2011, with exceptional drought and record-ing-setting temperatures in Texas and Oklahoma — and flooding in the northern Great Plains. Many locations in Texas and Oklahoma experienced more than 100 days over 100°F, with both states setting new high temperature records. Rates of water loss were double the long-term average, depleting water resources and contributing to more than $10 billion in direct losses to agriculture alone. In the future, average temperatures in this region are expected to increase and will continue to contribute to the intensity of heat waves.

By contrast, the Northern Plains were exceptionally wet, with Montana and Wyoming recording all-time wettest springs and the Dakotas and Nebraska not far behind. Record rainfall and snowmelt combined to push the Missouri River and its tributaries beyond their banks and leave much of the Crow Reservation in Montana underwater. The Souris River near Minot, North Dakota, crested at four feet above its previous record, causing losses estimated at $2 billion.

Projected climate change will have both positive and negative consequences for agricultural productivity in the Northern Plains, where increases in winter and spring precipitation will benefit productivity by increasing water availability through soil moisture reserves during the early growing season, but this can be offset by fields too wet to plant. Rising temperatures will lengthen the growing season, possibly allowing a second annual crop in some places and some years. However, warmer winters pose challenges.[5] Some pests and invasive weeds will be able to survive the warmer winters,[6] and winter crops that emerge from dormancy earlier are susceptible to spring freezes.[7]

In the Southern Plains, project-ed declines in precipitation in the south and greater evapora-tion everywhere due to higher temperatures will increase irri-gation demand and exacerbate current stresses on agricultural productivity. Increased water withdrawals from the Ogallala and High Plains Aquifers would accelerate ongoing depletion in the southern parts of the aquifers and limit the ability to irrigate.[8] Holding other aspects of production constant, the climate impacts of shifting from irrigated to dryland agriculture would reduce crop yields by about a factor of two.[9]

SELECTED RESPONSES

The Oglala Lakota tribe in South Dakota is incorporating climate change adaptation and mitigation planning as they consider long-term sustainable development. Their Oyate Omniciye plan is a partnership built around six livability principles related to transportation, housing, economic competitiveness, existing communities, federal investments, and local values. Their vision incorporates plans to reduce and adapt to future climate change while protecting cultural resources.[10]

SOUTHWEST

Snowpack and streamflow amounts are projected to decline in parts of the Southwest, decreasing surface water supply reliability for cities, agriculture, and ecosystems.

The Southwest produces more than half of the nation's high-value specialty crops, which are irrigation-dependent and particularly vulnerable to extremes of moisture, cold, and heat. Reduced yields from increasing temperatures and increasing competition for scarce water supplies will displace jobs in some rural communities.

Increased warming, drought, and insect outbreaks, all caused by or linked to climate change, have increased wildfires and impacts to people and ecosystems in the Southwest. Fire models project more wildfire and increased risks to communities across extensive areas.

Flooding and erosion in coastal areas are already occurring even at existing sea levels and damaging some California coastal areas during storms and extreme high tides. Sea level rise is projected to increase as Earth continues to warm, resulting in major damage as wind-driven waves ride upon higher seas and reach farther inland.

Projected regional temperature increases, combined with the way cities amplify heat, will pose increased threats and costs to public health in southwestern cities, which are home to more than 90% of the region's population. Disruptions to urban electricity and water supplies will exacerbate these health problems.

Heat, drought, and competition for water supplies will increase in the Southwest with continued climate change.

The Southwest is the hottest and driest region in the U.S., where the availability of water has defined its landscapes, history of human settlement, and modern economy. Climate changes pose challenges for an already parched region that is expected to get hotter and, in its southern half, significantly drier.

Increased heat and changes to rain and snowpack will send ripple effects throughout the region, affecting 56 million people – a population expected to increase to 94 million by 2050[1] – and its critical agriculture sector. Severe and sustained drought will stress water sources, already over-utilized in many areas, forcing increasing competition among farmers, energy producers, urban dwellers, and ecosystems for the region's most precious resource.

The region's populous coastal cities face rising sea levels, extreme high tides, and storm surges, which pose particular risks to highways, bridges, power plants, and sewage treatment plants. Climate-related challenges also increase risks to critical port cities, which handle half of the nation's incoming shipping containers. The region's rich diversity of plant and animal species will be increasingly stressed. Widespread tree death and fires, which already have caused billions of dollars in economic losses, are projected to increase. Tourism and recreation also face climate change challenges, including reduced streamflow and a shorter snow season, influencing everything from the ski industry to lake and river recreation.

Climate change contributes to increasing fires.

More than half of the nation's high-value specialty crops, including certain fruits, nuts, and vegetables, come from the Southwest. A longer frost-free season, less frequent cold air outbreaks, and more frequent heat waves accelerate crop ripening and maturity, reduce yields of corn, tree fruit, and wine grapes, stress livestock, and increase agricultural water consumption.[2] These changes are projected to continue and intensify, possibly requiring a northward shift in crop production, displacing existing growers and affecting farming communities.[3]

Winter chill periods are projected to fall below the duration necessary for many California trees to bear nuts and fruits, which will result in lower yields.[4]

Once temperatures increase beyond optimum growing thresholds, further increases, like those projected beyond 2050, can cause large decreases in crop yields and hurt the region's agricultural economy.

Longer Frost-Free Season Increases Stress on Crops

Graph shows significant increases in the number of consecutive frost-free days per year in the past three decades compared to the 1901-2010 average. This leads to further heat stress on plants and increased water demands for crops. Warmer winters can also lead to early bud burst or bloom of some perennial plants, resulting in frost damage when cold conditions occur in late spring. Higher winter temperatures also allow some agricultural pests to persist year round, and may allow new pests and diseases to become established.[14] (Figure source: Hoerling et al. 2013[10]).

Climate change is exacerbating the major factors that lead to wildfire: heat, drought, and dead trees.[5,6] Between 1970 and 2003, warmer and drier conditions increased burned area in western U.S. mid-elevation conifer forests by 650%.[7] Climate outweighed other factors in determining burned area in the western U.S. from 1916 to 2003.[8] Winter warming due to climate change has exacerbated bark beetle outbreaks by allowing more beetles, which normally die in cold weather, to survive and reproduce.[9] More wildfire is projected as climate change continues,[6,10,11,12] including a doubling of burned area in the southern Rockies,[11] and up to 74% more fires in California.[12] For more on fire in the Southwest see pages 53-54.

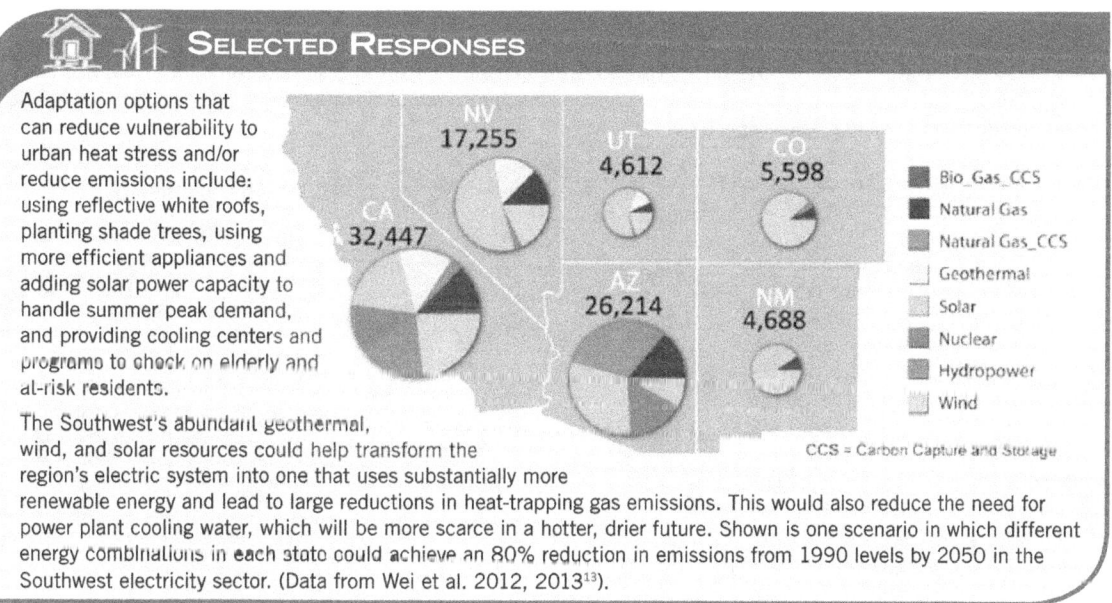

SELECTED RESPONSES

Adaptation options that can reduce vulnerability to urban heat stress and/or reduce emissions include: using reflective white roofs, planting shade trees, using more efficient appliances and adding solar power capacity to handle summer peak demand, and providing cooling centers and programs to check on elderly and at-risk residents.

The Southwest's abundant geothermal, wind, and solar resources could help transform the region's electric system into one that uses substantially more renewable energy and lead to large reductions in heat-trapping gas emissions. This would also reduce the need for power plant cooling water, which will be more scarce in a hotter, drier future. Shown is one scenario in which different energy combinations in each state could achieve an 80% reduction in emissions from 1990 levels by 2050 in the Southwest electricity sector. (Data from Wei et al. 2012, 2013[13]).

- Bio_Gas_CCS
- Natural Gas
- Natural Gas_CCS
- Geothermal
- Solar
- Nuclear
- Hydropower
- Wind

CCS = Carbon Capture and Storage

NV 17,255
UT 4,612
CO 5,598
CA 32,447
AZ 26,214
NM 4,688

NORTHWEST

KEY MESSAGES

Changes in the timing of streamflow related to changing snowmelt are already observed and will continue, reducing the supply of water for many competing demands and causing far-reaching ecological and socioeconomic consequences.

In the coastal zone, the effects of sea level rise, erosion, inundation, threats to infrastructure and habitat, and increasing ocean acidity collectively pose a major threat to the region.

The combined impacts of increasing wildfire, insect outbreaks, and tree diseases are already causing widespread tree die-off and are virtually certain to cause additional forest mortality by the 2040s and long-term transformation of forest landscapes. Under higher emissions scenarios, extensive conversion of subalpine forests to other forest types is projected by the 2080s.

While the agriculture sector's technical ability to adapt to changing conditions can offset some adverse impacts of a changing climate, there remain critical concerns for agriculture with respect to costs of adaptation, development of more climate resilient technologies and management, and availability and timing of water.

Rising summer temperatures and changing water flows threaten salmon and other fish species.

The Northwest's economy, infrastructure, natural systems, public health, and agriculture sectors all face important climate change related risks. Impacts on infrastructure, natural systems, human health, and economic sectors, combined with issues of social and ecological vulnerability, will unfold quite differently in largely natural areas, like the Cascade Range, than in urban areas like Seattle and Portland,[1] or among the region's many Native American tribes.[2]

Seasonal water patterns shape the life cycles of the region's flora and fauna, including iconic salmon and steelhead, and forested ecosystems.[3] Adding to the human influences on climate, human activities have altered natural habitats, threatened species, and extracted so much water that there are already conflicts among multiple users in dry years. As conflicts and trade-offs increase, the region's population continues to grow. Particularly in the face of climate change, the need to seek solutions to these conflicts is becoming increasingly urgent.

Observed regional warming has been linked to changes in the timing and amount of water availability in basins with significant snowmelt contributions to streamflow. By 2050, snowmelt is projected to shift three to four weeks earlier than the last century's average, and summer flows are projected to be substantially lower, even for a scenario that assumes emissions reductions (B1).[4] These reduced flows will require trade-offs among reservoir system objectives,[5] especially with the added challenges of summer increases in electric power demand for cooling and additional water consumption by crops and forests.

Future Shift in Timing of Streamflows

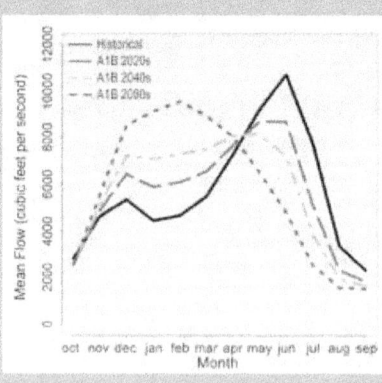

Mixed rain-snow watersheds, such as the Yakima River basin, an important agricultural area in eastern Washington, will see increased winter flows, earlier spring peak flows, and decreased summer flows in a warming climate, causing widespread impacts. Natural surface water availability during the already dry late summer period is projected to decrease across most of the Northwest.[6] Projections are based on the A1B emissions scenario, which assumes continued increases in emissions through mid century and gradual declines thereafter. (Figure source: adapted from Elsner et al. 2010[4]).

Insects and Fire in Northwest Forests

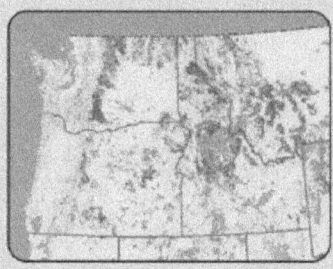
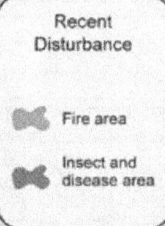
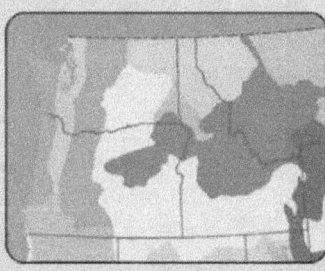

Recent Disturbance

Fire area

Insect and disease area

Projected Increase in Area Burned

600% to 700%
500% to 600%
400% to 500%
300% to 400%
200% to 300%
100% to 200%
Not modeled

(Left) Insects and fire have cumulatively affected large areas of the Northwest and are projected to be the dominant drivers of forest change in the near future. Map shows areas recently burned (1984 to 2008)[7] or affected by insects or disease (1997 to 2008).[8]
(Right) Map indicates the increases in area burned that would result from the regional temperature and precipitation changes associated with a 2.2°F global warming[9] across areas that share broad climatic and vegetation characteristics.[10] Local impacts will vary greatly within these broad areas with sensitivity of fuels to climate.[11]

Climate change will alter Northwest forests by increasing wildfire risk, insect and disease outbreaks, and by forcing longer-term shifts in forest types and species. Many impacts will be driven by water deficits, which increase tree stress and mortality, tree vulnerability to insects, and fuel flammability. By the 2080s, the median annual area burned in the Northwest would quadruple relative to the 1916-2007 period to 2 million acres (range 0.2 to 9.8 million acres) under a scenario that assumes continued increases in emissions through mid century and gradual declines thereafter (A1B).[11]

Oyster harvest in Coos Bay, Oregon. Ocean acidification poses threats to the region's important shellfish industry.

SELECTED ADAPTATION EFFORTS

In Washington's Nisqually River Delta, large-scale estuary restoration to assist salmon and wildlife recovery provides an example of adaptation to climate change and sea level rise. After a century of isolation behind dikes, much of the Nisqually National Wildlife Refuge was reconnected with tidal flow in 2009 by removal of a major dike and restoration of 762 acres, with the assistance of Ducks Unlimited and the Nisqually Indian Tribe. This reconnected more than 21 miles of historical tidal channels and floodplains with Puget Sound.[12] A new exterior dike was constructed to protect freshwater wetland habitat for migratory birds from tidal inundation, future sea level rise, and increasing river floods.

ALASKA

KEY MESSAGES

Arctic summer sea ice is receding faster than previously projected and is expected to virtually disappear before mid-century. This is altering marine ecosystems and leading to greater ship access, offshore development opportunity, and increased community vulnerability to coastal erosion.

Most glaciers in Alaska and British Columbia are shrinking substantially. This trend is expected to continue and has implications for hydropower production, ocean circulation patterns, fisheries, and global sea level rise.

Permafrost temperatures in Alaska are rising, a thawing trend that is expected to continue, causing multiple vulnerabilities through drier landscapes, more wildfire, altered wildlife habitat, increased cost of maintaining infrastructure, and the release of heat-trapping gases that increase climate warming.

Current and projected increases in Alaska's ocean temperatures and changes in ocean chemistry are expected to alter the distribution and productivity of Alaska's marine fisheries, which lead the U.S. in commercial value.

The cumulative effects of climate change in Alaska strongly affect Native communities, which are highly vulnerable to these rapid changes but have a deep cultural history of adapting to change.

Over the past 60 years, Alaska has warmed more than twice as rapidly as the rest of the U.S., with average annual air temperature increasing by 3°F and average winter temperature by 6°F, with substantial year-to-year and regional variability.[1] Most of the warming occurred around 1976 during a shift in a long-lived climate pattern (the Pacific Decadal Oscillation) from a cooler pattern to a warmer one. The underlying long-term warming trend has moderated the effects of the more recent shift of the Pacific Decadal Oscillation to its cooler phase in the early 2000s.[2] Alaska's warming involves more extremely hot days and fewer extremely cold days.[1,3] Because of its cold-adapted features and rapid warming, climate change impacts on Alaska are already pronounced, including earlier spring snowmelt, reduced sea ice, widespread glacier retreat, warmer permafrost, drier landscapes, and more extensive insect outbreaks and wildfire.

Inupiaq seal hunter on the Chukchi Sea. Reductions in sea ice alter food availability for many species from polar bear to walrus, and make hunting less safe for Alaska Native hunters.

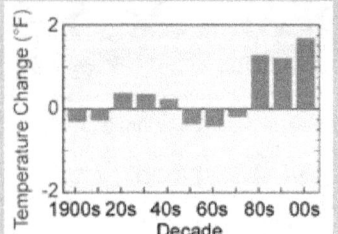

Rising Temperatures

Bars show Alaska average temperature changes by decade for 1901-2012 relative to the 1901-1960 average. The far right bar (2000s decade) includes 2011 and 2012. (Figure source: NOAA NCDC / CICS-NC).

The state's largest industries, energy production, mining, and fishing, are all affected by climate change. Continuing pressure for oil, gas, and mineral development on land and offshore in ice-covered waters increases the demand for infrastructure, placing additional stresses on ecosystems. Land-based energy exploration will be affected by a shorter season when ice roads are viable, yet reduced sea ice extent may create more opportunity for offshore development.

Alaska is home to 40% of the federally recognized tribes in the United States.[4] The small number of jobs, high cost of living, and rapid social change make rural, predominantly Native, communities highly vulnerable to climate change through impacts on traditional hunting and fishing and cultural connection to the land and sea.

Arctic sea ice extent and thickness have declined substantially, especially in late summer (September), when there is now only about half as much sea ice as at the beginning of the satellite record in 1979.[5,6] The seven Septembers with the lowest ice extent all occurred in the past seven years. Sea ice has also become thinner, with less ice lasting over multiple years, and is therefore more vulnerable to further melting.[6] Models that best match historical trends project that northern waters will be virtually ice-free in late summer by the 2030s.[7]

Reductions in sea ice increase the amount of the sun's energy absorbed by the ocean. This melts more ice, leaving more dark open water that gains even more heat, leading to a self-reinforcing cycle that increases warming.

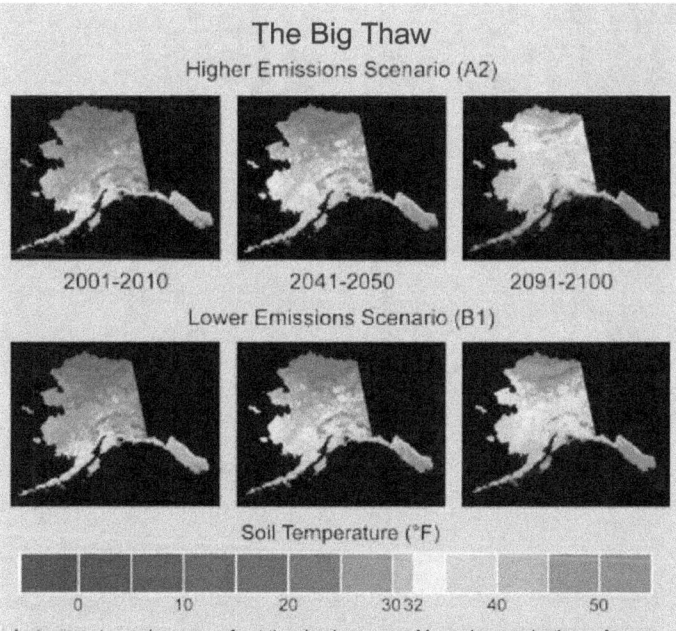

The Big Thaw
Higher Emissions Scenario (A2)

2001-2010   2041-2050   2091-2100

Lower Emissions Scenario (B1)

Soil Temperature (°F)

0   10   20   30 32   40   50

As temperatures rise, permafrost thawing increases. Maps show projections of average annual ground temperature at a depth of 3.3 feet for three time periods if emissions of heat-trapping gases continue to grow (higher scenario, A2), and if they are substantially reduced (lower scenario, B1). (Figure source: Permafrost Lab, Geophysical Institute, University of Alaska Fairbanks).

In Alaska, 80% of land is underlain by permafrost – frozen ground that restricts water drainage and therefore strongly influences landscape water balance and the design and maintenance of infrastructure. More than 70% of this area is vulnerable to subsidence (land sinking) upon thawing because of its ice content.[8] Permafrost near the Alaskan Arctic coast has warmed 6°F to 8°F at 3.3 foot depth since the mid-1980s.[9] Thawing is already occurring in interior and southern Alaska, where permafrost temperatures are near the thaw point.[10] Permafrost will continue to thaw,[11] and some models project that near-surface permafrost will be lost entirely from large parts of Alaska by the end of this century.[12]

SELECTED RESPONSES

Local governments and tribes throughout Alaska are planting native vegetation, moving inland or away from rivers, and building riprap walls, seawalls, or groins, which are shore-protection structures built perpendicular to the shoreline.[13] Top photo shows a Homer seawall battered by waves while still under construction.

Villages including Newtok, Shishmaref (bottom), and Kivalina are facing relocation because of sea level rise and coastal erosion. Storm surges that used to be buffered by ice are now causing more shoreline and infrastructure damage. Residents of these villages face thawing permafrost, tilting houses, and sinking boardwalks along with aging fuel tanks and other infrastructure. Newtok has worked for a generation to move to a safer location, but current federal legislation does not authorize federal or state agencies to assist communities in relocating, or the use of public funds to repair or upgrade storm-damaged infrastructure in flood prone locations.[14] Shishmaref and Kivalina are also seeking to relocate but have been similarly unsuccessful.

HAWAI'I AND PACIFIC ISLANDS

KEY MESSAGES

Warmer oceans are leading to increased coral bleaching events and disease outbreaks in coral reefs, as well as changed distribution patterns of tuna fisheries. Ocean acidification will reduce coral growth and health. Warming and acidification, combined with existing stresses, will strongly affect coral reef fish communities.

Freshwater supplies are already constrained and will become more limited on many islands. Saltwater intrusion associated with sea level rise will reduce the quantity and quality of freshwater in coastal aquifers, especially on low islands. In areas where precipitation does not increase, freshwater supplies will be adversely affected as air temperature rises.

Increasing temperatures, and in some areas reduced rainfall, will stress native Pacific Island plants and animals, especially in high-elevation ecosystems with increasing exposure to invasive species, increasing the risk of extinctions.

Rising sea levels, coupled with high water levels caused by storms, will incrementally increase coastal flooding and erosion, damaging coastal ecosystems, infrastructure, and agriculture, and negatively affecting tourism.

Mounting threats to food and water security, infrastructure, health, and safety are expected to lead to increasing human migration, making it increasingly difficult for Pacific Islanders to sustain the region's many unique customs, beliefs, and languages.

The U.S. Pacific Islands are at risk from climate changes that will affect nearly every aspect of life. The region includes more than 2,000 islands spanning millions of square miles of ocean. Rising air and ocean temperatures, shifting rainfall patterns, changing frequencies and intensities of storms and drought, decreasing streamflows, rising sea levels, and changing ocean chemistry will threaten the sustainability of globally important and diverse ecosystems on land and in the oceans, as well as local communities, livelihoods, and cultures.

On most islands, increased temperatures coupled with decreased rainfall and increased drought will reduce the amount of freshwater available for drinking and crop irrigation.[1] Climate change impacts on freshwater resources will vary with differing island size and topography, affecting water storage capability and susceptibility to coastal flooding. Low-lying islands will be particularly vulnerable due to their small land mass, geographic isolation, limited potable water sources, and limited agricultural resources.[2] Sea level rise will increase saltwater intrusion from the ocean during storms.[3,4]

Rising sea levels will escalate the threat to coastal structures and property, groundwater reservoirs, harbor operations, airports, wastewater systems, shallow coral reefs, sea grass beds, intertidal flats and mangrove forests, and other social, economic, and natural resources.

"High" and "Low" Pacific Islands Face Different Threats

The Pacific Islands include "high" volcanic islands, such as that on the left, that reach nearly 14,000 feet above sea level, and "low" atolls and islands, such as that on the right, that peak at just a few feet above present sea level. (Left) Ko'olau Mountains on the windward side of Oahu, Hawai'i. (Right) Laysan Island, Papahānaumokuākea Marine National Monument.

Coastal infrastructure and agricultural activity on low islands will be affected as sea level rise decreases the land area available for farming,[3] and periodic flooding increases the salinity of groundwater.

Many of Hawai'i's native birds, marvels of evolution largely limited to high-elevation forests, are increasingly vulnerable as rising temperatures allow mosquitoes carrying diseases like avian malaria to thrive at higher elevations.[5] Mangrove area in the region could decline 10% to 20% in this century due to sea level rise.[6] This would reduce the nursery areas, feeding grounds, and habitat for fish, crustaceans, and other species, as well as shoreline protection and wave dampening, and water

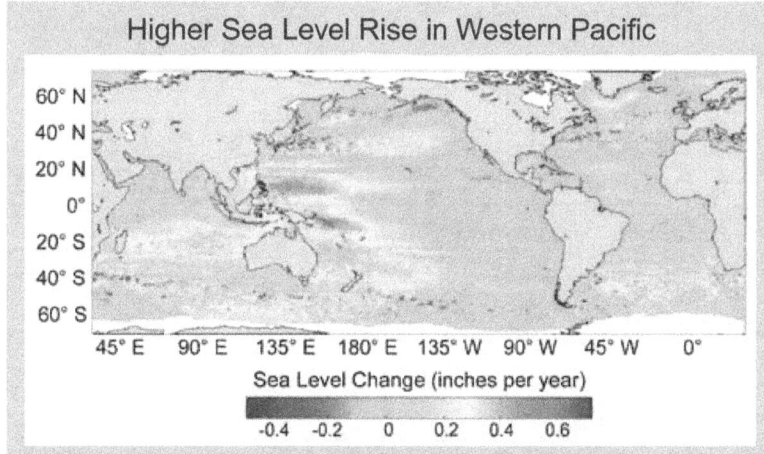

Higher Sea Level Rise in Western Pacific

Sea Level Change (inches per year)

-0.4 -0.2   0   0.2   0.4   0.6

Map shows large variations across the Pacific Ocean in sea level trends for 1993-2010. The largest sea level increase has been observed in the Western Pacific, due, in part to changing wind patterns associated with natural climate variability. (Figure source: adapted from Merrifield 2011[11] by permission of American Meteorological Society).

filtration provided by mangroves.[7] Pacific seabirds that breed on low-lying atolls will lose large portions of their breeding populations[8] as their habitat is increasingly and more extensively covered by seawater.

Economic impacts from tourism loss will be greatest on islands with more developed infrastructure. In Hawai'i, for example, where tourism comprises 26% of the state's economy, damage to tourism infrastructure could have large economic impacts – the loss of Waikīkī Beach alone could lead to an annual loss of $2 billion in visitor expenditures.[9]

Because Pacific Islands are almost entirely dependent upon imported food, fuel, and material, the vulnerability of ports and airports to extreme events, sea level rise, and increasing wave heights is of great concern. Climate change is also expected to have serious effects on human health, for example by increasing the incidence of dengue fever.[10] In addition, sea level rise and flooding are expected to overwhelm sewer systems and threaten public sanitation.

The traditional lifestyles and cultures of Indigenous communities in all Pacific Islands will be seriously affected by climate change. Drought threatens traditional food sources such as taro and breadfruit, and coral death from warming-induced bleaching will threaten subsistence fisheries in island communities.[4] Climate change impacts, coupled with socioeconomic or political motivations, may be great enough to lead some people to relocate. Depending on the scale and distance of migration, a variety of challenges face migrants and the communities receiving them.

Increasing ocean temperature and acidity threaten coral reef ecosystems. By 2100, assuming ongoing increases in emissions of heat-trapping gases (A2 scenario), continued loss of coral reefs and the shelter they provide will result in extensive losses in numbers and species of reef fishes.[12] For more on ocean impacts, see pages 59-60.

SELECTED ADAPTATION

The State of Hawai'i, in cooperation with university, private, state, and federal scientists and others, has drafted an adaptation plan,[13] one of the priorities of which is preserving water sources through conservation of the forests, as indicated in their "Rain Follows The Forest" report.[14]

RURAL COMMUNITIES

KEY MESSAGES

Rural communities are highly dependent upon natural resources for their livelihoods and social structures. Climate change related impacts are currently affecting rural communities. These impacts will progressively increase over this century and will shift the locations where rural economic activities (like agriculture, forestry, and recreation) can thrive.

Rural communities face particular geographic and demographic obstacles in responding to and preparing for climate change risks. In particular, physical isolation, limited economic diversity, and higher poverty rates, combined with an aging population, increase the vulnerability of rural communities. Systems of fundamental importance to rural populations are already stressed by remoteness and limited access.

Responding to additional challenges from climate change impacts will require significant adaptation within rural transportation and infrastructure systems, as well as health and emergency response systems. Governments in rural communities have limited institutional capacity to respond to, plan for, and anticipate climate change impacts.

More than 95% of U.S. land area is classified as rural, but is home to just 19% of the population.[1] Rural areas provide natural resources that much of the rest of the U.S. depends on for food, energy, water, forests, recreation, national character, and quality of life.[2] Rural economic foundations and community cohesion are intricately linked to these natural systems, which are inherently vulnerable to climate change. Urban areas that depend on goods and services from rural areas will also be affected by climate change driven impacts across the countryside.

Warming, climate volatility, extreme weather events, and environmental change are already affecting the economies and cultures of rural areas. Many communities face considerable risk to their infrastructure, livelihoods, and quality of life from observed and projected climate shifts. These changes will progressively increase volatility in food commodity markets, shift locations where particular economic activities can thrive, alter the ranges of plant and animal species, and, depending on the region, increase water scarcity, exacerbate flooding and coastal erosion, and increase the intensity and frequency of wildfires across the rural landscape. Because many rural communities are less diverse than urban areas in their economic activities, changes in the viability of one traditional economic sector will place disproportionate stresses on community stability.

Rural America has already experienced impacts of climate change related weather effects, including crop and livestock loss from severe drought and flooding,[3] damage to levees and roads from extreme storms,[4] shifts in planting and harvesting times,[5] and large-scale losses from fires and other weather-related disasters.[6] These impacts have profound effects, often significantly affecting the health and well-being of rural residents and communities, and are amplified by the essential economic link between these communities and their natural resource base.

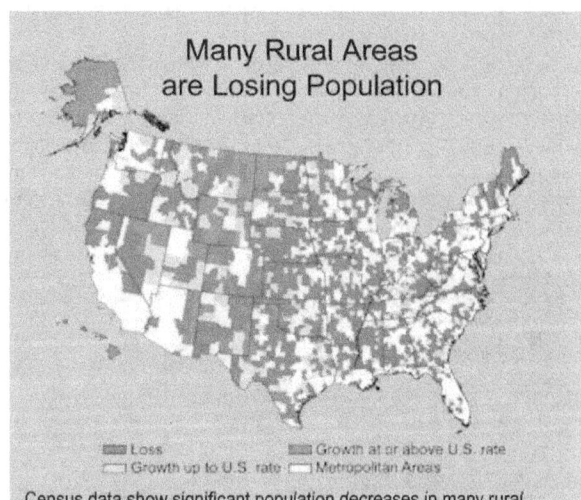

Census data show significant population decreases in many rural areas, notably in the Great Plains. Many rural communities' existing vulnerabilities to climate change, including physical isolation, reduced services like health care, and an aging population, are projected to increase as population decreases. (Figure source: USDA Economic Research Service 2013[7]).

Hunting, fishing, bird watching, and other wildlife-related activities will be affected as wildlife habitats shift and relationships among species change.[8] Cold-weather rec-reation and tourism will be adversely affected by climate change. Snow accumulation in the West has decreased, and is expected to continue to de-crease, as a result of observed

Flooded corn field and river flood waters illustrate threats rural areas face in a changing climate.

and projected warming. Similar changes to snowpack are expected in the Northeast.[9] Adverse impacts on winter sports are projected to be more pronounced in the Northeast and Southwest.[10]

Coastal areas will be adversely affected by sea level rise and increased severity of storms.[11] Changing conditions, such as wetland loss and beach erosion in coastal areas,[12] and increased risk of natural hazards such as wildfire, flash flooding, storm surge, river flooding, drought, and extremely high temperatures can alter the character and attraction of rural areas as tourist destinations.

Changing demographics and economic activities influence the ability to respond to climate change. Rural areas are char-acterized by higher unemployment, more dependence on government transfer payments, less diversified economies, and fewer social and economic resources needed for resilience in the face of climate change.[10,13]

 ## ADAPTATION CHALLENGES

Climate variability and increases in temperature, extreme events (such as storms, floods, heat waves, and droughts), and sea level rise are expected to have widespread impacts on the provision of services from state, regional, local, and tribal governments. Emergency management, energy use and distribution systems, transportation and infrastructure planning, and public health will all be affected.

Rural governments often depend heavily on volunteers to meet community challenges like fire protection or flood response. Rural communities have limited locally available financial resources to cope with the effects of climate change. Small community size tends to make services expensive or available only by traveling some distance.

Adaptation efforts require planning, but local governance structures tend to de-emphasize planning capacity compared to urban areas. While 73% of metropolitan counties have land-use planners, only 29% of rural counties not adjacent to a metropolitan county had one or more planners. Moreover, rural communities are not equipped to deal with major infrastructure expenses.[14]

If rural communities are to respond adequately to future climate changes, they will likely need help assessing their risks and vulnerabilities, prioritizing and coordinating projects, funding and allocating financial and human resources, and deploying information-sharing and decision support tools.

Impacts due to climate change will cross community and regional lines, making solutions dependent upon meaningful participation of numerous stakeholders from federal, state, local, and tribal governments, science and academia, the private sector, non-profit organizations, and the general public. Effective adaptation measures are closely tied to specific local conditions and needs and take into account existing social networks.[15]

Decisions regarding adaptation responses for both urban and rural populations can occur at various scales (federal, state, local, tribal, private sector, and individual) but need to take interdependencies into account. Many decisions that significantly affect rural communities may not be under the control of local governments or rural residents.

Timing is a critical aspect of adaptation and mitigation, so engaging rural residents early in decision processes about investments in public infrastructure, protection of shorelines, changes in insurance provision, or new management initiatives can influence behavior and choices in ways that enhance positive outcomes of adaptation and mitigation.

COASTS

Coastal lifelines, such as water supply and energy infrastructure and evacuation routes, are increasingly vulnerable to higher sea levels and storm surges, inland flooding, erosion, and other climate-related changes.

Nationally important assets, such as ports, tourism, and fishing sites, in already-vulnerable coastal locations, are increasingly exposed to sea level rise and related hazards. This threatens to disrupt economic activity within coastal areas and the regions they serve and results in significant costs from protecting or moving these assets.

Socioeconomic disparities create uneven exposures and sensitivities to growing coastal risks and limit adaptation options for some coastal communities, resulting in the displacement of the most vulnerable people from coastal areas.

Coastal ecosystems are particularly vulnerable to climate change because many have already been dramatically altered by human stresses; climate change will result in further reduction or loss of the services that these ecosystems provide, including potentially irreversible impacts.

Leaders and residents of coastal regions are increasingly aware of the high vulnerability of coasts to climate change, and are developing plans to prepare for potential impacts on citizens, businesses, and environmental assets. Significant institutional, political, social, and economic obstacles to implementing adaptation actions remain.

More than 50% of Americans – 164 million people – live in coastal counties, with 1.2 million added each year. Residents, combined with the more than 180 million tourists that flock to the coasts each year,[1,2] place heavy demands on the unique natural systems and resources that make coastal areas so attractive and productive.[1,2]

No other region concentrates so many people and so much economic activity on so little land, while also being so relentlessly affected by the sometimes violent interactions of land, sea, and air. Humans have heavily altered the coastal environment through development, changes in land use, and overexploitation of resources.

Now, the changing climate is imposing additional stresses,[3] making life on the coast more challenging. The consequences will ripple through the entire nation.

Damage to coastal roads is already a problem along the shores of the U.S. and will worsen as sea level continues to rise.

Paths of Hurricanes Katrina and Rita Relative to Oil and Gas Production Facilities

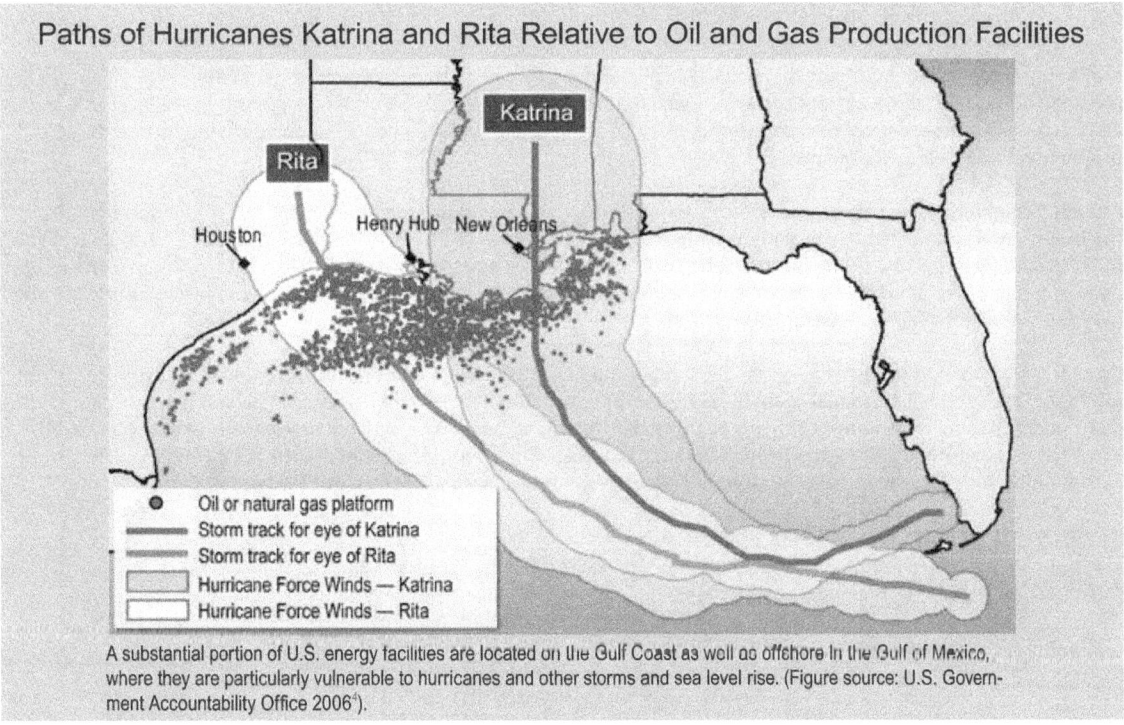

Legend:
- ● Oil or natural gas platform
- ── Storm track for eye of Katrina
- ⋯ Storm track for eye of Rita
- ▨ Hurricane Force Winds — Katrina
- ▢ Hurricane Force Winds — Rita

A substantial portion of U.S. energy facilities are located on the Gulf Coast as well as offshore in the Gulf of Mexico, where they are particularly vulnerable to hurricanes and other storms and sea level rise. (Figure source: U.S. Government Accountability Office 2006[4]).

Lifelines at Risk

Key coastal vulnerabilities arise from complex interactions among climate change and other physical, human, and ecological factors. These vulnerabilities have the potential to fundamentally alter life at the coast and disrupt coast-dependent economic activities.

The more than 60,000 miles of coastal roads are essential for human activities. Already, many coastal roads are affected during storm events[5] and extreme high tides.[6] As coastal bridges, tunnels, and roads are built or redesigned, engineers must account for present and future climate change impacts.[7]

Wastewater management and drainage systems are also at risk. Systems will become overwhelmed with increased rainfall intensity over more impervious surfaces, such as asphalt and concrete.[8] Sea level rise will cause a variety of problems including salt water intrusion into coastal aquifers.[9] Together, climate change impacts increase the risks of urban flooding, combined sewer overflows, deteriorating coastal water quality, and human health impacts.[10]

The nation's energy infrastructure, such as power plants, oil and gas refineries, storage tanks, transformers, and electricity transmission lines, are often located directly in the coastal floodplain.[11] Roughly two-thirds of imported oil enters the U.S. through Gulf of Mexico ports,[12] and unless adaptive measures are taken, storm-related flooding, erosion, and permanent inundation from sea level rise will disrupt the supply of refined products to the rest of the nation.[13]

There are a variety of options to protect, replace, and redesign existing infrastructure, including flood proofing and flood protection through dikes, berms, pumps, integration of natural landscape features, elevation, more frequent upgrades, or relocation.[14] Such adaptation options are best assessed in a site-specific context, weighing social, economic, and ecological considerations.

Natural gas platform in the Gulf of Mexico illustrates some of the infrastructure at risk from coastal storms.

Economic Disruption

More than 5,790 square miles and more than $1 trillion of property and structures are at risk of inundation from sea level rise of two feet above current sea level – which could be reached by 2050 under a high rate of sea level rise, by 2070 assuming a lower rate of rise, and sooner in areas of rapid land subsidence.[15,16,17] Roughly half of the vulnerable property value is located in Florida.[16,18]

Although comprehensive national estimates are not yet available, regional studies are indicative of the potential risk: the incremental annual damage of climate change to capital assets in the Gulf region alone could be $2.7 to $4.6 billion by 2030, and $8.3 to $13.2 billion by 2050; about 20% of these at-risk assets are in the oil and gas industry.[19] Investing approximately $50 billion for adaptation over the next 20 years could lead to approximately $135 billion in averted losses over the lifetime of adaptive measures.[19,20]

Coastal recreation and tourism comprises the largest and fastest-growing sector of the U.S. service industry, accounting for 85% of the $700 billion annual tourism-related revenues.[1,21] Hard shoreline protection against the encroaching sea (like building sea walls or riprap) generally aggravates erosion and beach loss, and causes negative effects on coastal ecosystems, undermining the attractiveness of beach tourism. Thus, "soft protection," such as beach replenishment or conservation and restoration of sand dunes and wetlands, is increasingly preferred to "hard protection" measures.

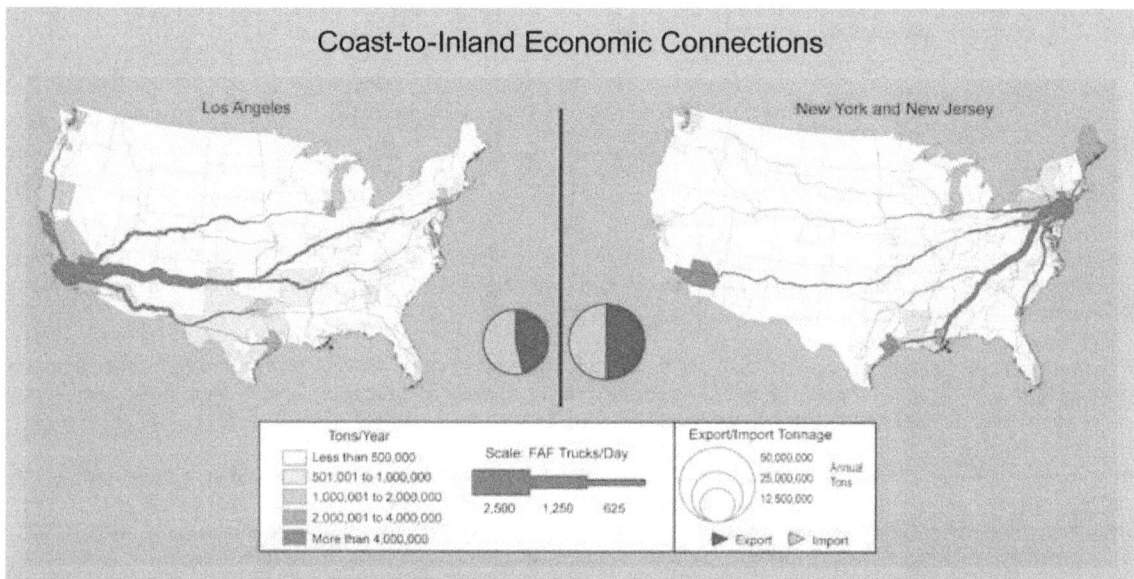

Ports are deeply interconnected with inland areas through the goods imported and exported each year. Climate change impacts on ports can thus have far-reaching implications for the nation's economy. Maps show the exports and imports in 2010 (in tons/year) and freight flows (in trucks per day) from two major U.S. ports (Los Angeles and New York/New Jersey) to other U.S. areas designated in the U.S. Department of Transportation's Freight Analysis Framework (FAF). Note: Highway Link Flow less than 5 FAF Trucks/Day are not shown. (Figure source: U.S. Department of Transportation, Federal Highway Administration, Office of Freight Management and Operations, Freight Analysis Framework, version 3.4, 2012[22]).

Socioeconomic Disparities

There are large socioeconomic disparities in coastal areas,[23,24] and a full understanding of risk for coastal communities requires consideration of social vulnerability factors that limit people's ability to adapt. These factors include lower income, minority status, low educational achievement, advanced age, lower economic and social mobility, and much lower likelihood of being insured than wealthy property owners.[25] The most socially vulnerable populations also tend to have fewer adaptation options in their current locations, and thus may be at greater risk of dislocation.[24,26]

Vulnerable Ecosystems

Coastal ecosystems provide a suite of valuable benefits (ecosystem services) on which humans depend, including reducing the impacts from floods, buffering from storm surge and waves, and providing nursery habitat for important fish and other species, water filtration, carbon storage, and opportunities for recreation and enjoyment.[27,28]

However, many of these ecosystems and the services they provide are rapidly being degraded by human impacts, including pollution, habitat destruction, and the spread of invasive species.

These existing stresses on coastal ecosystems will be exacerbated by climate change effects, such as increased ocean temperatures that lead to coral bleaching,[29] altered river flows affecting the health of estuaries,[30] and acidified waters threatening shellfish.[31] Of particular concern is the potential for coastal ecosystems to cross thresholds of rapid change ("tipping points"), beyond which they exist in a dramatically altered state or are lost entirely from the area. Some ecosystems are already near tipping points and in some cases the changes will be irreversible.[32]

ADAPTATION CHALLENGES AND OPPORTUNITIES

Coastal leaders and populations are increasingly concerned about climate-related impacts and are developing adaptation plans,[33] but support for development restrictions or managed retreat is limited.[34]

Enacting measures that increase resilience in the face of current hazards, while reducing long-term risks due to climate change, continues to be challenging.[35]

A robust finding is that the cost of inaction is 4 to 10 times greater than the cost associated with preventive hazard mitigation.[16,36] Even so, prioritizing expenditures now whose benefits accrue far in the future is difficult.[37]

Cumulative costs to the economy of responding to sea level rise and flooding events alone could be as high as $325 billion by 2100 for 4 feet of sea level rise, with $130 billion expected to be incurred in Florida and $88 billion in the North Atlantic region.[17] The projected costs associated with one foot of sea level rise by 2100 are roughly

A coastal ecosystem restoration project in New York City integrates revegetation (a form of green infrastructure) with bulkheads and rip-rap (gray or built infrastructure). Investments in coastal ecosystem conservation and restoration can protect coastal waterfronts and infrastructure, while providing additional benefits, such as habitat for commercial and recreational fish, birds, and other animal and plant species, that are not offered by built infrastructure.

$200 billion. These figures exclude losses of valuable ecosystem services, as well as indirect losses from business disruption, lost economic activity, impacts on economic growth, or other non-market losses.[17,38]

Property insurance can serve as an important mode of financial adaptation to climate risks,[39] but the full potential of leveraging insurance rates and availability has not yet been realized.[40,41] Federal fiscal exposure for the National Flood Insurance Program was estimated at nearly $1.3 trillion in 2012.[42] Reforms were enacted in 2012, though various challenges remain.[43]

Climate adaptation efforts that integrate hazard mitigation, natural resource conservation, and restoration of coastal ecosystems can enhance ecological resilience and reduce the exposure of property, infrastructure, and economic activities to climate change impacts.[28,44] Yet, the integration and translation of scientific understanding of the benefits provided by ecosystems into engineering design and hazard management remains challenging.[45] Adaptation efforts to date that have begun to connect these issues across jurisdictional and departmental boundaries and create innovative solutions are thus extremely encouraging.[40,46]

ALASKA

- Summer sea ice is receding rapidly, altering marine ecosystems, allowing for greater ship access and offshore development, and making Native communities highly susceptible to coastal erosion.

- Ice loss from melting Alaskan and Canadian glaciers currently contributes almost as much to sea level rise as does melting of the Greenland Ice Sheet.

- Current and projected increases in Alaska's ocean temperatures and changes in ocean chemistry are expected to alter the distribution and productivity of Alaska's marine fisheries.

NORTHWEST

- The substantial global sea level rise is regionally moderated by the continuing uplift of land, with few exceptions, such as the Seattle area and central Oregon.

- Commercial shellfish populations are at risk from ocean acidification.

- The region's relatively high economic dependence on commercial fisheries makes it sensitive to climate change impacts on marine species and ecosystems and related coastal ecosystems.

- Coastal storm surges are expected to be higher due to increases in sea level alone, and more intense persistent storm tracks (atmospheric river systems) will increase coastal flooding risks from inland runoff.

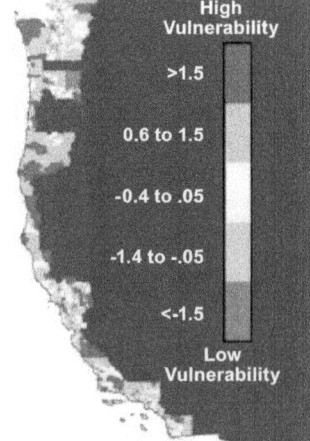

High
Vulnerability

>1.5

0.6 to 1.5

-0.4 to .05

-1.4 to -.05

<-1.5

Low
Vulnerability

CALIFORNIA

- Sea level has risen approximately 7 inches from 1900 to 2005, and is expected to rise at growing rates in this century.

- Higher temperatures; changes in precipitation, runoff, and water supplies; and saltwater intrusion into coastal aquifers will result in negative impacts on coastal water resources.

- Coastal storm surges are expected to be higher due to increases in sea level alone, and more intense persistent storm tracks (atmospheric river systems) will increase coastal flooding risks from inland runoff.

- Expensive coastal development, critical infrastructure, and valuable coastal wetlands are at growing risk from coastal erosion, temporary flooding, and permanent inundation.

- The San Francisco Bay and San Joaquin/Sacramento River Delta are particularly vulnerable to sea level rise and changes in salinity, temperature, and runoff; endangering one of the ecological "jewels" of the West Coast, as well as growing development, and crucial water infrastructure.

HAWAI'I AND PACIFIC ISLANDS

- Warmer and drier conditions will reduce freshwater supplies on many Pacific Islands, especially on low lying islands and atolls.

- Sea level rise will continue at accelerating rates, exacerbating coastal erosion, damaging infrastructure and agriculture, reducing critical habitat, and threatening shallow coral reef systems.

- Extreme water levels occur when high tides combine with interannual and interdecadal sea level variations (such as El Niño Southern Oscillation, Pacific Decadal Oscillation, mesoscale eddy events) and storm surge.

- Coral reef changes pose threats to communities, cultures, and ecosystems.

Boxes summarize coastal climate change threats for each region.
Map shows how social vulnerability varies around the coasts.[48]

GREAT LAKES

- Higher temperatures and longer growing seasons in the Great Lakes region favor production of blue-green and toxic algae that can harm fish, water quality, habitat, and aesthetics.

- Increased winter air temperatures will lead to decreased Great Lakes ice cover, making shorelines more susceptible to erosion and flooding.

- Current projections of lake level changes are uncertain.

NORTHEAST

- Highly built-up coastal corridor concentrates population and supporting infrastructure.

- Storm surges from nor'easters and hurricanes can cause significant damage.

- The historical rate of relative sea level rise varies across the region.

- Wetlands and estuaries are vulnerable to inundation from sea level rise; buildings and infrastructure are most vulnerable to higher storm surges as sea level rises.

SOCIAL VULNERABILITY

Map shows a Social Vulnerability Index, providing a quantitative, integrative measure of vulnerability of human populations in the U.S. High vulnerability (dark pink) typically indicates some combination of high exposure and high sensitivity to the effects of climate change and low capacity to deal with them. Index components and weighting are specific to each region (North Atlantic, South Atlantic, Gulf, Pacific, Great Lakes, Alaska, and Hawai'i), and are constructed from Census data including measures of poverty, age, family structure, location (rural versus urban), foreign-born status, wealth, gender, Native American status, and occupation.[24,47]

MID-ATLANTIC

- Rates of local sea level rise in the Chesapeake Bay are greater than the global average.

- Sea level rise and related flooding and erosion threaten coastal homes, infrastructure, and commercial development, including ports.

- Chesapeake Bay ecosystems are already heavily degraded, making them more vulnerable to climate-related impacts.

SOUTHEAST AND CARIBBEAN

- A large number of cities, critical infrastructure, and water supplies are at low elevations and exposed to sea level rise, in some places moderated by land uplift.

- Ecosystems of the Southeast are vulnerable to loss from relative sea level rise, especially tidal marshes and swamps.

- Sea level rise will affect coastal agriculture through higher storm surges, saltwater intrusion, and impacts on freshwater supplies.

- The number of land-falling tropical storms may decline, reducing important rainfall.

- The incidence of harmful algal blooms is expected to increase with climate change, as are health problems previously uncommon in the region.

GULF COAST

- Hurricanes, land subsidence, sea level rise, and erosion already pose great risks to Gulf Coast areas, placing homes, critical infrastructure, and people at risk, and causing permanent land loss.

- Coastal inland and water temperatures are expected to rise; coastal inland areas are expected to become drier.

- There is still uncertainty about future frequency and intensity of Gulf of Mexico hurricanes, but sea level rise will increase storm surges.

- The Florida Keys, South Florida, and coastal Louisiana are particularly vulnerable to additional sea level rise and saltwater intrusion.

FUTURE
NATIONAL CLIMATE ASSESSMENTS

Sustained Assessment

Since 1990, Congress has required periodic updates on climate science and its implications. A primary goal of the National Climate Assessment (NCA) is to help the nation anticipate, mitigate, and adapt to impacts from climate change in the context of other national and global change factors.

As this third NCA was being prepared, a vision for a new approach to assessments took shape. This vision includes an ongoing process for understanding and evaluating the nation's vulnerabilities to climate change and its capacity to respond. A sustained assessment, in addition to producing quadrennial assessment reports as required by law, recognizes that the ability to understand, predict, assess, and respond to rapid changes in the global environment requires ongoing efforts to integrate new knowledge and experience.

A sustained assessment process would: 1) advance the science needed to improve the assessment process and its outcomes, building associated foundational knowledge and collecting relevant data; 2) develop targeted scientific reports and other products that respond directly to the needs of federal agencies, state and local governments, tribes, and other decision-makers; 3) create a framework for continued interactions between the assessment partners and stakeholders and the scientific community; and 4) support the capacity of those engaged in assessment activities to maintain such interactions.

To provide decision-makers with more timely, concise, and useful information, a sustained assessment process would include both ongoing, extensive engagement with public and private partners and targeted, scientifically rigorous reports that address concerns in a timely fashion. A growing

A sustained assessment process would provide decision-makers with more timely and useful information.

body of assessment literature has guided and informed the development of this approach to a sustained assessment.[1] The envisioned sustained assessment process includes continuing and expanding engagement with scientists and other professionals from government, academia, business, and non-governmental organizations. These partnerships broaden the knowledge base from which conclusions can be drawn. In addition, sustained engagement with decision-makers and end users helps scientists understand what information society wants and needs, and provides mechanisms for researchers to receive ongoing feedback on the utility of the tools and data they provide.

An ongoing process that supports these forms of outreach and engagement allows for more comprehensive and insightful evaluation of climate changes across the nation, including how decision-makers and end users are responding to these changes. The most thoughtful and robust responses to climate change can be made only when these complex issues, including the underlying science and its many implications for the nation, are documented and communicated in a way that both scientists and non-scientists can understand. This sustained assessment process will lead to better outcomes by providing more relevant, comprehensible, and usable knowledge to guide decisions related to climate change at local, regional, and national scales. More information is available in the NCADAC special report "Preparing the Nation for Change: Building a Sustained National Climate Assessment."[2]

Ongoing monitoring and observations can help guide decision-making.

U.S. GLOBAL CHANGE RESEARCH PROGRAM

In addition to producing the quadrennial assessment reports required by the 1990 Global Change Research Act, a well-designed and executed sustained assessment process would produce many other important outcomes:

1. Increase the nation's capacity to measure and evaluate the impacts of and responses to further climate change in the U.S., locally, regionally, and nationally.
2. Improve the collection of assessment-related critical data, access to those data, and the capacity of users to work with datasets – including their use in decision support tools – relevant to their specific issues and interests. This includes periodically assessing how users are applying such data.
3. Support the creation of the first integrated suite of national indicators of climate-related trends across a variety of important climate drivers and responses.
4. Catalyze the production of targeted, in-depth reports on various topics that help inform choices about mitigation and adaptation. These reports would generate new insights about climate change, its impacts, and the effectiveness of societal responses. In addition, other reports could focus on improvements to aspects of the process (for example, scenarios and indicators) to reinforce the foundation for the quadrennial assessments.
5. Facilitate, support, and leverage a network of scientific, decision-maker, and user communities for extended dialog and engagement regarding climate change.
6. Provide a systematic way to identify gaps in knowledge and uncertainties faced by the scientific community and by U.S. domestic and international partners and to assist in setting priorities for their resolution.
7. Enhance integration with other assessment efforts such as the Intergovernmental Panel on Climate Change and modeling efforts such as the Coupled Model Intercomparison Project.
8. Develop and apply tools to evaluate progress and guide improvements in processes and products over time, supporting an iterative approach to managing risks and opportunities associated with changing conditions.

Research Needs

Five priority research goals have been identified to advance future climate and global change assessments.

- Improve understanding of the climate system and its drivers.
- Improve understanding of climate impacts and vulnerability.
- Increase understanding of adaptation pathways.
- Identify the mitigation options that reduce the risk of longer-term climate change.
- Improve decision support and integrated assessment.

This assessment also identifies five cross-cutting foundational capabilities that are essential for advancing the ability to continue to conduct climate and global change assessments and for addressing the five research goals.

- Integrate natural and social science, engineering, and other disciplinary approaches.
- Ensure availability of observations, monitoring, and infrastructure for critical data collection and analysis.
- Build capacity for climate assessment through training, education, and workforce development.
- Enhance the development and use of scenarios.
- Promote international research and collaboration.

These are not intended to prescribe a specific research agenda but rather to summarize the research needs and gaps that emerged during development of this NCA that are relevant to the development of future research plans.

For example, several important topics could not be comprehensively covered in this assessment and could be considered in future reports. These include analyses of the economic costs of climate change impacts (and the associated benefits of mitigation and adaptation strategies); the considerations related to climate change for U.S. national security, as appropriate, as a topic integrated with other regional and sectoral discussions; and the interactions of adaptation and mitigation options, including consideration of the co-benefits and potential unintended consequences of particular decisions.

The following criteria should be considered in establishing research priorities that support assessments:

- Promote understanding of the fundamental behavior of the Earth's climate and environmental systems.
- Promote understanding of the socioeconomic impacts of a changing climate.
- Build capacity to assess risks and consequences.
- Support research that enables the infrastructure needed for analysis.
- Build decision support capacity.
- Support engagement of the private sector and investment communities.
- Leverage private sector, university, and international resources and partnerships.

CONCLUDING THOUGHTS

As climate change and its impacts become more prevalent, Americans face choices. Although some additional climate change and related impacts are now unavoidable, the amount of future climate change and its consequences will still largely be determined by our choices, now and in the near future. There is still time to act to limit the amount of climate change and the extent of damaging impacts we will face.

This report offers an overview of some of the options and activities being implemented or planned around the

> There is still time to act to limit the amount of climate change and the extent of damaging impacts.

country as governments, businesses, and individuals begin to respond to climate change. These include efforts to reduce heat-trapping emissions and adapt to changing conditions.

There are many pathways to significantly reduce heat-trapping gas emissions. In addition, actions to reduce emissions can yield benefits for objectives apart from managing climate change, such as increasing energy security and improving human health. Similarly, actions to prepare for and adapt to climate change impacts can also improve our resilience in other ways.

Across the nation, Americans are beginning to act:

Managing Heavy Rainfall

Municipalities across the country are increasingly implementing a range of adaptation options to manage the increase in heavy downpours, including using green roofs, rain gardens, roadside plantings, porous pavement, and rainwater harvesting. These techniques typically utilize soils and vegetation to absorb runoff close to where it falls, limiting flooding and sewer backups. In Maine, an initiative is underway to help towns adapt culverts to handle the heavier rainfalls already occurring and expected to increase further over the lifetime of the culverts. People are creating decision tools to map culvert locations, schedule maintenance, estimate needed culvert size, and analyze replacement needs and costs. There are complex, multi-jurisdictional challenges for even such seemingly simple actions as using larger culverts to carry water from major storms.

Cities Mitigate and Adapt

Many cities are undertaking initiatives to reduce heat-trapping gas emissions. More than 1,055 municipalities from all 50 states have signed the U.S. Mayors Climate Protection Agreement, and many of these communities are actively implementing strategies to reduce their greenhouse gas footprint. By integrating climate-change considerations into daily operations, some cities are forestalling the need to develop new or isolated climate change specific policies or procedures. This strategy enables cities and other government agencies to take advantage of existing funding sources and programs and achieve co-benefits in areas such as sustainability, public health, economic development, disaster preparedness, and environmental justice. Pursuing low-cost, no-regrets options is a particularly attractive short-term strategy for many cities.

U.S. GLOBAL CHANGE RESEARCH PROGRAM

Achieving the lower emissions pathway used in this assessment would require substantial decarbonization of the global economy by the end of this century, implying a fundamental transformation of the global energy system.

Many technologies are potentially available to accomplish emissions reduction. They include ways to increase the efficiency of energy use and facilitate a shift to low-carbon energy sources, improvements in the cost and performance of renewables (such as wind, solar, and bioenergy) and nuclear energy, ways to reduce the cost of carbon capture and storage, means to expand carbon sinks through management of forests and soils and increased agricultural productivity, and phasing down the use of

Climate change presents us with both challenges and opportunities.

other heat-trapping gases, like hydrofluorocarbons (HFCs), widely used for refrigeration.

The United States has declared a goal of reducing its greenhouse gas emissions about 17% below 2005 levels by 2020 through a range of actions, including limiting carbon emissions from power plants and continuing to increase the fuel economy of cars and trucks and the energy efficiency of buildings. The U.S. has also indicated that it will seek to exert leadership internationally.

Climate change presents us with both challenges and opportunities. The information contained in this report can help enable our society to effectively respond and prepare for our future.

Northeast Takes Action

The most well-known, multi-state effort has been the Regional Greenhouse Gas Initiative (RGGI), formed by ten northeastern and Mid-Atlantic states (though New Jersey exited in 2011). RGGI is a cap-and-trade system applied to the power sector with revenue from allowance auctions directed to investments in efficiency and renewable energy.

California Acts to Reduce Emissions

California's Global Warming Solutions Act (AB 32) is an ambitious law that sets a state goal to reduce its greenhouse gas emissions to 1990 levels by 2020. The state program caps emissions and uses a market-based system of trading in emissions credits (cap-and-trade), limits imports of baseload electricity generation from coal and oil, and implements a number of other regulatory actions.

Southwest Ramps Up Renewables

The Southwest's abundant geothermal, wind, and solar power-generation resources could help transform the region's electric generating system into one that uses substantially more renewable energy. This transformation has already started, driven in part by renewable energy portfolio standards that require a certain amount of electricity to be generated with renewables. These standards have been adopted by five of six Southwest states, and also include renewable energy goals in Utah.

1. Overview and Report Findings
Convening Lead Authors
Jerry Melillo, Marine Biological Laboratory
Terese (T.C.) Richmond, Van Ness Feldman, LLP
Gary Yohe, Wesleyan University

2. Our Changing Climate
Convening Lead Authors
John Walsh, University of Alaska Fairbanks
Donald Wuebbles, University of Illinois
Lead Authors
Katharine Hayhoe, Texas Tech University
James Kossin, NOAA National Climatic Data Center
Kenneth Kunkel, CICS-NC, North Carolina State Univ.,
 NOAA National Climatic Data Center
Graeme Stephens, NASA Jet Propulsion Laboratory
Peter Thorne, Nansen Environmental and Remote Sensing Center
Russell Vose, NOAA National Climatic Data Center
Michael Wehner, Lawrence Berkeley National Laboratory
Josh Willis, NASA Jet Propulsion Laboratory
Contributing Authors
David Anderson, NOAA National Climatic Data Center
Scott Doney, Woods Hole Oceanographic Institution
Richard Feely, NOAA Pacific Marine Environmental Laboratory
Paula Hennon, CICS-NC, North Carolina State Univ.,
 NOAA National Climatic Data Center
Viatcheslav Kharin, Canadian Centre for Climate Modelling and Analysis,
 Environment Canada
Thomas Knutson, NOAA Geophysical Fluid Dynamics Laboratory
Felix Landerer, NASA Jet Propulsion Laboratory
Tim Lenton, Exeter University
John Kennedy, UK Meteorological Office
Richard Somerville, Scripps Institution of Oceanography,
 Univ. of California, San Diego

3. Water Resources
Convening Lead Authors:
Aris Georgakakos, Georgia Institute of Technology
Paul Fleming, Seattle Public Utilities
Lead Authors:
Michael Dettinger, U.S. Geological Survey
Christa Peters-Lidard, National Aeronautics and Space Administration
Terese (T.C.) Richmond, Van Ness Feldman, LLP
Ken Reckhow, Duke University
Kathleen White, U.S. Army Corps of Engineers
David Yates, University Corporation for Atmospheric Research

4. Energy Supply and Use
Convening Lead Authors
Jan Dell, ConocoPhillips
Susan Tierney, Analysis Group Consultants
Lead Authors
Guido Franco, California Energy Commission
Richard G. Newell, Duke University
Rich Richels, Electric Power Research Institute
John Weyant, Stanford University
Thomas J. Wilbanks, Oak Ridge National Laboratory

5. Transportation
Convening Lead Authors
Henry G. Schwartz, HGS Consulting, LLC
Michael Meyer, Parsons Brinckerhoff
Lead Authors
Cynthia J. Burbank, Parsons Brinckerhoff
Michael Kuby, Arizona State University
Clinton Oster, Indiana University
John Posey, East-West Gateway Council of Governments
Edmond J Russo, U.S. Army Corps of Engineers
Arthur Rypinski, U.S. Department of Transportation

6. Agriculture
Convening Lead Authors
Jerry Hatfield, U.S. Department of Agriculture
Gene Takle, Iowa State University
Lead Authors
Richard Grotjahn, University of California, Davis
Patrick Holden, Waterborne Environmental, Inc.
R. Cesar Izaurralde, Pacific Northwest National Laboratory
Terry Mader, University of Nebraska, Lincoln
Elizabeth Marshall, U.S. Department of Agriculture
Contributing Authors
Diana Liverman, University of Arizona

7. Forests
Convening Lead Authors
Linda A. Joyce, U.S. Forest Service
Steven W. Running, University of Montana
Lead Authors
David D. Breshears, University of Arizona
Virginia H. Dale, Oak Ridge National Laboratory
Robert W. Malmsheimer, SUNY Environmental Science and Forestry
R. Neil Sampson, Vision Forestry, LLC
Brent Sohngen, Ohio State University
Christopher W. Woodall, U.S. Forest Service

8. Ecosystems, Biodiversity, and Ecosystem Services

Convening Lead Authors

Peter M. Groffman, Cary Institute of Ecosystem Studies

Peter Kareiva, The Nature Conservancy

Lead Authors

Shawn Carter, U.S. Geological Survey

Nancy B. Grimm, Arizona State University

Josh Lawler, University of Washington

Michelle Mack, University of Florida

Virginia Matzek, Santa Clara University

Heather Tallis, Stanford University

9. Human Health

Convening Lead Authors

George Luber, Centers for Disease Control and Prevention

Kim Knowlton, Natural Resources Defense Council and Mailman School of
  Public Health, Columbia University

Lead Authors

John Balbus, National Institutes of Health

Howard Frumkin, University of Washington

Mary Hayden, National Center for Atmospheric Research

Jeremy Hess, Emory University

Michael McGeehin, RTI International

Nicky Sheats, Thomas Edison State College

Contributing Authors

Lorraine Backer, Centers for Disease Control and Prevention

C. Ben Beard, Centers for Disease Control and Prevention

Kristie L. Ebi, ClimAdapt, LLC

Edward Maibach, George Mason University

Richard S. Ostfeld, Cary Institute of Ecosystem Studies

Christine Wiedinmyer, National Center for Atmospheric Research

Emily Zielinski-Gutiérrez, Centers for Disease Control and Prevention

Lewis Ziska, U.S. Department of Agriculture

10. Energy, Water, and Land Use

Convening Lead Authors

Kathy Hibbard, Pacific Northwest National Laboratory

Tom Wilson, Electric Power Research Institute

Lead Authors

Kristen Averyt, University of Colorado Boulder

Robert Harriss, Environmental Defense Fund

Robin Newmark, National Renewable Energy Laboratory

Steven Rose, Electric Power Research Institute

Elena Shevliakova, Princeton University

Vincent Tidwell, Sandia National Laboratories

11. Urban Systems, Infrastructure, and Vulnerability

Convening Lead Authors

Susan L. Cutter, University of South Carolina

William Solecki, City University of New York

Lead Authors

Nancy Bragado, City of San Diego

JoAnn Carmin, Massachusetts Institute of Technology

Michail Fragkias, Boise State University

Matthias Ruth, Northeastern University

Thomas J. Wilbanks, Oak Ridge National Laboratory

12. Indigenous Peoples, Lands, and Resources

Convening Lead Authors

T.M. Bull Bennett, Kiksapa Consulting, LLC

Nancy G. Maynard, National Aeronautics and Space Administration and
  University of Miami

Lead Authors

Patricia Cochran, Alaska Native Science Commission

Robert Gough, Intertribal Council on Utility Policy

Kathy Lynn, University of Oregon

Julie Maldonado, American University,
  University Corporation for Atmospheric Research

Garrit Voggesser, National Wildlife Federation

Susan Wotkyns, Northern Arizona University

Contributing Authors

Karen Cozzetto, University of Colorado at Boulder

13. Land Use and Land Cover Change

Convening Lead Authors

Daniel G. Brown, University of Michigan

Colin Polsky, Clark University

Lead Authors

Paul Bolstad, University of Minnesota

Samuel D. Brody, Texas A&M University at Galveston

David Hulse, University of Oregon

Roger Kroh, Mid-America Regional Council

Thomas R. Loveland, U.S. Geological Survey

Allison Thomson, Pacific Northwest National Laboratory

14. Rural Communities

Convening Lead Authors

David Hales, Second Nature

William Hohenstein, U.S. Department of Agriculture

Lead Authors

Marcie D. Bidwell, Mountain Studies Institute

Craig Landry, East Carolina University

David McGranahan, U.S. Department of Agriculture

Joseph Molnar, Auburn University

Lois Wright Morton, Iowa State University

Marcela Vasquez, University of Arizona

Contributing Authors

Jenna Jadin, U.S. Department of Agriculture

15. Biogeochemical Cycles

Convening Lead Authors

James N. Galloway, University of Virginia

William H. Schlesinger, Cary Institute of Ecosystem Studies

Lead Authors

Christopher M. Clark, U.S. Environmental Protection Agency

Nancy B. Grimm, Arizona State University

Robert B. Jackson, Duke University

Beverly E. Law, Oregon State University

Peter E. Thornton, Oak Ridge National Laboratory

Alan R. Townsend, University of Colorado Boulder

Contributing Author

Rebecca Martin, Washington State University Vancouver

16. Northeast

Convening Lead Authors

Radley Horton, Columbia University

Gary Yohe, Wesleyan University

Lead Authors

William Easterling, Pennsylvania State University

Robert Kates, University of Maine

Matthias Ruth, Northeastern University

Edna Sussman, Fordham University School of Law

Adam Whelchel, The Nature Conservancy

David Wolfe, Cornell University

Contributing Author

Fredric Lipschultz, NASA and Bermuda Institute of Ocean Sciences

17. Southeast and Caribbean

Convening Lead Authors

Lynne M. Carter, Louisiana State University

James W. Jones, University of Florida

Lead Authors

Leonard Berry, Florida Atlantic University

Virginia Burkett, U.S. Geological Survey

James F. Murley, South Florida Regional Planning Council

Jayantha Obeysekera, South Florida Water Management District

Paul J. Schramm, Centers for Disease Control and Prevention

David Wear, U.S. Forest Service

18. Midwest

Convening Lead Authors

Sara C. Pryor, Indiana University

Donald Scavia, University of Michigan

Lead Authors

Charles Downer, U.S. Army Engineer Research and Development Center

Marc Gaden, Great Lakes Fishery Commission

Louis Iverson, U.S. Forest Service

Rolf Nordstrom, Great Plains Institute

Jonathan Patz, University of Wisconsin

G. Philip Robertson, Michigan State University

19. Great Plains

Convening Lead Authors

Mark Shafer, Oklahoma Climatological Survey

Dennis Ojima, Colorado State University

Lead Authors

John M. Antle, Oregon State University

Doug Kluck, National Oceanic and Atmospheric Administration

Renee A. McPherson, University of Oklahoma

Sascha Petersen, Adaptation International

Bridget Scanlon, University of Texas

Kathleen Sherman, Colorado State University

20. Southwest

Convening Lead Authors

Gregg Garfin, University of Arizona

Guido Franco, California Energy Commission

Lead Authors

Hilda Blanco, University of Southern California

Andrew Comrie, University of Arizona

Patrick Gonzalez, National Park Service

Thomas Piechota, University of Nevada, Las Vegas

Rebecca Smyth, National Oceanic and Atmospheric Administration

Reagan Waskom, Colorado State University

21. Northwest

Convening Lead Authors
Philip Mote, Oregon State University
Amy K. Snover, University of Washington

Lead Authors
Susan Capalbo, Oregon State University
Sanford D. Eigenbrode, University of Idaho
Patty Glick, National Wildlife Federation
Jeremy Littell, U.S. Geological Survey
Richard Raymondi, Idaho Department of Water Resources
Spencer Reeder, Cascadia Consulting Group

22. Alaska

Convening Lead Authors
F. Stuart Chapin III, University of Alaska Fairbanks
Sarah F. Trainor, University of Alaska Fairbanks

Lead Authors
Patricia Cochran, Alaska Native Science Commission
Henry Huntington, Huntington Consulting
Carl Markon, U.S. Geological Survey
Molly McCammon, Alaska Ocean Observing System
A. David McGuire, U.S. Geological Survey and University of Alaska Fairbanks
Mark Serreze, University of Colorado

23. Hawai'i and U.S. Affiliated Pacific Islands

Convening Lead Authors
Jo-Ann Leong, University of Hawai'i
John J. Marra, National Oceanic and Atmospheric Administration

Lead Authors
Melissa L. Finucane, East-West Center
Thomas Giambelluca, University of Hawai'i
Mark Merrifield, University of Hawai'i
Stephen E. Miller, U.S. Fish and Wildlife Service
Jeffrey Polovina, National Oceanic and Atmospheric Administration
Eileen Shea, National Oceanic and Atmospheric Administration

Contributing Authors
Maxine Burkett, University of Hawai'i
John Campbell, University of Waikato
Penehuro Lefale, Meteorological Service of New Zealand Ltd.
Fredric Lipschultz, NASA and Bermuda Institute of Ocean Sciences
Lloyd Loope, U.S. Geological Survey
Deanna Spooner, Pacific Island Climate Change Cooperative
Bin Wang, University of Hawai'i

24. Oceans and Marine Resources

Convening Lead Authors
Scott Doney, Woods Hole Oceanographic Institution
Andrew A. Rosenberg, Union of Concerned Scientists

Lead Authors
Michael Alexander, National Oceanic and Atmospheric Administration
Francisco Chavez, Monterey Bay Aquarium Research Institute
C. Drew Harvell, Cornell University
Gretchen Hofmann, University of California Santa Barbara
Michael Orbach, Duke University
Mary Ruckelshaus, Natural Capital Project

25. Coastal Zone Development and Ecosystems

Convening Lead Authors
Susanne C. Moser, Susanne Moser Research & Consulting and
  Stanford University
Margaret A. Davidson, National Oceanic and Atmospheric Administration

Lead Authors
Paul Kirshen, University of New Hampshire
Peter Mulvaney, Skidmore, Owings & Merrill LLP
James F. Murley, South Florida Regional Planning Council
James E. Neumann, Industrial Economics, Inc.
Laura Petes, National Oceanic and Atmospheric Administration
Denise Reed, The Water Institute of the Gulf

26. Decision Support: Connecting Science, Risk Perception, and Decisions

Convening Lead Authors
Richard Moss, Joint Global Change Research Institute,
  Pacific Northwest National Laboratory, University of Maryland
P. Lynn Scarlett, The Nature Conservancy

Lead Authors
Melissa A. Kenney, University of Maryland
Howard Kunreuther, University of Pennsylvania
Robert Lempert, RAND Corporation
Jay Manning, Cascadia Law Group
B. Ken Williams, The Wildlife Society

Contributing Authors
James W. Boyd, Resources for the Future
Emily T. Cloyd, University Corporation for Atmospheric Research
Laurna Kaatz, Denver Water
Lindene Patton, Zurich North America

27. Mitigation

Convening Lead Authors

Henry D. Jacoby, Massachusetts Institute of Technology

Anthony C. Janetos, Boston University

Lead Authors

Richard Birdsey, U.S. Forest Service

James Buizer, University of Arizona

Katherine Calvin, Pacific Northwest National Laboratory, University of Maryland

Francisco de la Chesnaye, Electric Power Research Institute

David Schimel, NASA Jet Propulsion Laboratory

Ian Sue Wing, Boston University

Contributing Authors

Reid Detchon, United Nations Foundation

Jae Edmonds, Pacific Northwest National Laboratory, University of Maryland

Lynn Russell, Scripps Institution of Oceanography,
  University of California, San Diego

Jason West, University of North Carolina

28. Adaptation

Convening Lead Authors

Rosina Bierbaum, University of Michigan

Arthur Lee, Chevron Corporation

Joel Smith, Stratus Consulting

Lead Authors

Maria Blair, Independent

Lynne M. Carter, Louisiana State University

F. Stuart Chapin III, University of Alaska Fairbanks

Paul Fleming, Seattle Public Utilities

Susan Ruffo, The Nature Conservancy

Contributing Authors

Shannon McNeeley, Colorado State University

Missy Stults, University of Michigan

Laura Verduzco, Chevron Corporation

Emily Seyller, University Corporation for Atmospheric Research

29. Research Needs for Climate and Global Change Assessments

Convening Lead Authors

Robert W. Corell, Florida International University and the GETF Center for
  Energy and Climate Solutions

Diana Liverman, University of Arizona

Lead Authors

Kirstin Dow, University of South Carolina

Kristie L. Ebi, ClimAdapt, LLC

Kenneth Kunkel, CICS-NC, North Carolina State Univ.,
  NOAA National Climatic Data Center

Linda O. Mearns, National Center for Atmospheric Research

Jerry Melillo, Marine Biological Laboratory

30. Sustained Assessment: A New Vision for Future U.S. Assessments

Convening Lead Authors

John A. Hall, U.S. Department of Defense

Maria Blair, Independent

Lead Authors

James L. Buizer, University of Arizona

David I. Gustafson, Monsanto Company

Brian Holland, ICLEI – Local Governments for Sustainability

Susanne C. Moser, Susanne Moser Research & Consulting and
  Stanford University

Anne M. Waple, Second Nature and University Corporation for
  Atmospheric Research

Appendix 3. Climate Science Supplement, and Appendix 4. Frequently Asked Questions

Convening Lead Authors

John Walsh, University of Alaska Fairbanks

Donald Wuebbles, University of Illinois

Lead Authors

Katharine Hayhoe, Texas Tech University

James Kossin, NOAA National Climatic Data Center

Kenneth Kunkel, CICS-NC, North Carolina State Univ.,
  NOAA National Climatic Data Center

Graeme Stephens, NASA Jet Propulsion Laboratory

Peter Thorne, Nansen Environmental and Remote Sensing Center

Russell Vose, NOAA National Climatic Data Center

Michael Wehner, Lawrence Berkeley National Laboratory

Josh Willis, NASA Jet Propulsion Laboratory

Contributing Authors

David Anderson, NOAA National Climatic Data Center

Viatcheslav Kharin, Canadian Centre for Climate Modelling and Analysis,
  Environment Canada

Thomas Knutson, NOAA Geophysical Fluid Dynamics Laboratory

Felix Landerer, NASA Jet Propulsion Laboratory

Tim Lenton, Exeter University

John Kennedy, UK Meteorological Office

Richard Somerville, Scripps Institution of Oceanography,
  Univ. of California, San Diego

USGCRP National Climate Assessment Coordination Office

Katharine Jacobs, Director, National Climate Assessment, White House Office of Science and Technology Policy (OSTP) (through December 2013) / University of Arizona

Fabien Laurier, Director, Third National Climate Assessment, White House OSTP (previously Deputy Director, USGCRP) (from December 2013)

Glynis Lough, NCA Chief of Staff, USGCRP / UCAR (from June 2012)

Sheila O'Brien, NCA Chief of Staff, USGCRP / UCAR (through May 2012)

Susan Aragon-Long, NCA Senior Scientist and Sector Coordinator, U.S. Geological Survey

Ralph Cantral, NCA Senior Scientist and Sector Coordinator, NOAA (through November 2012)

Tess Carter, Student Assistant, Brown University

Emily Therese Cloyd, NCA Public Participation and Engagement Coordinator, USGCRP / UCAR

Chelsea Combest-Friedman, NCA International Coordinator, Knauss Marine Policy Fellow, NOAA (February 2011-February 2012)

Alison Delgado, NCA Scientist and Sector Coordinator, Pacific Northwest National Laboratory, Joint Global Change Research Institute, University of Maryland (from October 2012)

William Emanuel, NCA Senior Scientist and Sector Coordinator, Pacific Northwest National Laboratory, Joint Global Change Research Institute, University of Maryland (June 2011-September 2012)

Matt Erickson, Student Assistant, Washington State University (July-October 2012)

Ilya Fischhoff, NCA Program Coordinator, USGCRP / UCAR

Elizabeth Fly, NCA Coastal Coordinator, Knauss Marine Policy Fellow, NOAA (February 2013-January 2014)

Chelcy Ford, NCA Sector Coordinator, USFS (August-November 2011)

Wyatt Freeman, Student Assistant, George Mason University / UCAR (May-September 2012)

Bryce Golden-Chen, NCA Program Coordinator, USGCRP / UCAR

Nancy Grimm, NCA Senior Scientist and Sector Coordinator, NSF / Arizona State University (July 2011-September 2012)

Tess Hart, NCA Communications Assistant, USGCRP / UCAR (June-July 2011)

Melissa Kenney, NCA Indicators Coordinator, NOAA / University of Maryland

Fredric Lipschultz, NCA Senior Scientist and Regional Coordinator, NASA / Bermuda Institute of Ocean Sciences

Stuart Luther, Student Assistant, Arizona State University / UCAR (June-August 2011)

Julie Maldonado, NCA Engagement Assistant and Tribal Coordinator, USGCRP / UCAR

Krista Mantsch, Student Assistant, Indiana University / UCAR (May-September 2013)

Rebecca Martin, Student Assistant, Washington State University (June-August 2012)

Paul Schramm, NCA Sector Coordinator, Centers for Disease Control and Prevention (June-November 2010)

Technical Support Unit, National Climatic Data Center, NOAA/NESDIS

David Easterling, NCA Technical Support Unit Director, NOAA National Climatic Data Center (from March 2013)

Anne Waple, NCA Technical Support Unit Director, NOAA NCDC / UCAR (through February 2013)

Susan Joy Hassol, Senior Science Writer, Climate Communication, LLC / Cooperative Institute for Climate and Satellites, North Carolina State University (CICS-NC)

Paula Ann Hennon, NCA Technical Support Unit Deputy Director, CICS-NC

Kenneth Kunkel, Chief Scientist, CICS-NC

Sara W. Veasey, Creative Director, NOAA NCDC

Andrew Buddenberg, Software Engineer/Scientific Programmer, CICS-NC

Fred Burnett, Administrative Assistant, Jamison Professional Services, Inc.

Sarah Champion, Scientific Data Curator and Process Analyst, CICS-NC

Doreen DiCarlo, Program Coordinator, CICS-NC (August 2011-April 2012)

Daniel Glick, Editor, CICS-NC

Jessicca Griffin, Lead Graphic Designer, CICS-NC

John Keck, Web Consultant, LMI, Inc. (August 2010 - September 2011)

Angel Li, Web Developer, CICS-NC

Clark Lind, Administrative Assistant, The Baldwin Group, Inc. (January-September 2012)

Liz Love-Brotak, Graphic Designer, NOAA NCDC

Tom Maycock, Technical Editor, CICS-NC

Janice Mills, Business Manager, CICS-NC

Deb Misch, Graphic Designer, Jamison Professional Services, Inc.

Julie Moore, Administrative Assistant, The Baldwin Group, Inc. (June 2010-January 2012)

Ana Pinheiro-Privette, Data Coordinator, CICS-NC (January 2012-July 2013)

Deborah B. Riddle, Graphic Designer, NOAA NCDC

April Sides, Web Developer, ERT, Inc.

Laura E. Stevens, Research Scientist, CICS-NC

Scott Stevens, Support Scientist, CICS-NC

Brooke Stewart, Science Editor/Production Coordinator, CICS-NC

Liqiang Sun, Research Scientist/Modeling Support, CICS-NC

Robert Taylor, Student Assistant, UNC Asheville, CICS-NC

Devin Thomas, Metadata Specialist, ERT, Inc.

Teresa Young, Print Specialist, Team ERT/STG, Inc.

Review Editors

Joseph Arvai, University of Calgary

Peter Backlund, University Corporation for Atmospheric Research

Lawrence Band, University of North Carolina

Jill S. Baron, U.S. Geological Survey / Colorado State University

Michelle L. Bell, Yale University

Donald Boesch, University of Maryland

Joel R. Brown, New Mexico State University

Ingrid C. (Indy) Burke, University of Wyoming

Gina Campoli, Vermont Agency of Transportation

Mary Anne Carroll, University of Michigan

Scott L. Collins, University of New Mexico

John Daigle, University of Maine
Ruth DeFries, Columbia University
Lisa Dilling, University of Colorado
Otto C. Doering III, Purdue University
Hadi Dowlatabadi, University of British Columbia
Charles T. Driscoll, Syracuse University
Hallie C. Eakin, Arizona State University
John Farrington, Woods Hole Oceanographic Institution
Chris E. Forest, Pennsylvania State University
Efi Foufoula-Georgiou, University of Minnesota
Adam Freed, The Nature Conservancy
Robert Fri, Resources for the Future
Stephen T. Gray, U.S. Geological Survey
Jay Gulledge, Oak Ridge National Laboratory
Terrie Klinger, University of Washington
Ian Kraucunas, Pacific Northwest National Laboratory
Larissa Larsen, University of Michigan
William J. Massman, U.S. Forest Service
Michael D. Mastrandrea, Stanford University
Pamela Matson, Stanford University
Ronald G. Prinn, Massachusetts Institute of Technology
J.C. Randolph, Indiana University
G. Philip Robertson, Michigan State University
David Robinson, Rutgers University
Dork Sahagian, Lehigh University
Christopher A. Scott, University of Arizona
Peter Vitousek, Stanford University
Andrew C. Wood, NOAA

United States Global Change Research Program
Thomas Armstrong (OSTP), Executive Director, USGCRP
Chris Weaver (OSTP / EPA), Deputy Executive Director, USGCRP

Subcommittee on Global Change Research
Chair
Thomas Karl, U.S. Department of Commerce
Vice Chairs
Ann Bartuska, U.S. Department of Agriculture, Vice Chair, Adaptation Science
Gerald Geernaert, U.S. Department of Energy, Vice Chair, Integrated Modeling
Mike Freilich, National Aeronautics and Space Administration, Vice Chair, Integrated Observations
Roger Wakimoto, National Science Foundation, Vice-Chair
Principals
John Balbus, U.S. Department of Health and Human Services
Katharine Batten, U.S. Agency for International Development
Joel Clement, U.S. Department of the Interior
Robert Detrick, U.S. Department of Commerce
Scott L. Harper, U.S. Department of Defense
Leonard Hirsch, Smithsonian Institution
William Hohenstein, U.S. Department of Agriculture

Jack Kaye, National Aeronautics and Space Administration
Michael Kuperberg, U.S. Department of Energy
C. Andrew Miller, U.S. Environmental Protection Agency
Arthur Rypinski, U.S. Department of Transportation
Joann Roskoski, National Science Foundation
Trigg Talley, U.S. Department of State

Interagency National Climate Assessment Working Group
Chair
Katharine Jacobs, White House Office of Science and Technology Policy (through December 2013)
Fabien Laurier, White House Office of Science and Technology Policy (from December 2013)
Vice-Chair
Virginia Burkett, U.S. Department of the Interior – U.S. Geological Survey (from March 2013)
Anne Waple, NOAA NCDC / UCAR (through February 2013)

National Aeronautics and Space Administration
Allison Leidner, Earth Science Division / Universities Space Research Association

National Science Foundation
Anjuli Bamzai, Directorate for Geosciences (through May 2011)
Eve Gruntfest, Directorate for Geosciences (January-November 2013)
Rita Teutonico, Directorate for Social, Behavioral, and Economic Sciences (through January 2011)

Smithsonian Institution
Leonard Hirsch, Office of the Undersecretary for Science

U.S. Department of Agriculture
Linda Langner, U.S. Forest Service (through January 2011)
Carolyn Olson, Office of the Chief Economist
Toral Patel-Weynand, U.S. Forest Service
Louie Tupas, National Institute of Food and Agriculture
Margaret Walsh, Office of the Chief Economist

U.S. Department of Commerce
Ko Barrett, National Oceanic and Atmospheric Administration (from February 2013)
David Easterling, National Oceanic and Atmospheric Administration – National Climatic Data Center (from March 2013)
Nancy McNabb, National Institute of Standards and Technology (from February 2013)
Adam Parris, National Oceanic and Atmospheric Administration
Anne Waple, NOAA NCDC / UCAR (through February 2013)

U.S. Department of Defense

William Goran, U.S. Army Corps of Engineers
John Hall, Office of the Secretary of Defense
Katherine Nixon, Navy Task Force Climate Change (from May 2013)
Courtney St. John, Navy Task Force Climate Change (through August 2012)

U.S. Department of Energy

Robert Vallario, Office of Science

U.S. Department of Health and Human Services

John Balbus, National Institutes of Health
Paul Schramm, Centers for Disease Control and Prevention (through July 2011)

U.S. Department of Homeland Security

Mike Kangior, Office of Policy (from November 2011)
John Laws, National Protection and Programs Directorate (from May 2013)

U.S. Department of the Interior

Susan Aragon-Long, U.S. Geological Survey
Virginia Burkett, U.S. Geological Survey
Leigh Welling, National Park Service (through May 2011)

U.S. Department of State

David Reidmiller, Bureau of Oceans and International Environmental
 & Scientific Affairs
Kenli Kim, Bureau of Oceans and International Environmental
 & Scientific Affairs (from February 2013)

U.S. Department of Transportation

Arthur Rypinski, Office of the Secretary
Mike Savonis, Federal Highway Administration (through March 2011)
AJ Singletary, Office of the Secretary (through August 2010)

U.S. Environmental Protection Agency

Rona Birnbaum, Office of Air and Radiation
Anne Grambsch, Office of Research and Development
Lesley Jantarasami, Office of Air and Radiation

White House Council on Environmental Quality

Jeff Peterson (through July 2013)
Jamie Pool (from February 2013)

White House Office of Management and Budget

Stuart Levenbach (through May 2012)

White House Office of Science and Technology Policy

Katharine Jacobs, Environment and Energy Division (through December 2013)
Fabien Laurier, Environment and Energy Division (from December 2013)

With special thanks to former NOAA Administrator, Jane Lubchenco and former
Associate Director of the Office of Science and Technology Policy, Shere Abbott

pg. 33–Coral bleaching: courtesy Ernesto Weil; Farmer observing drought: ©Scott Olson/Getty Images

pg. 34–Satellite image of smoke and fires: courtesy NASA/GSFC

pg. 35–Person sneezing: ©Jose Luis Pelaez, Inc./Blend Images/Corbis

pg. 36–Man wiping forehead: ©Richard Drew/AP/Corbis

pg. 38–Utility worker: ©Gene Blevins; Worker inspecting damaged road: ©AP Photo/The Virginian-Pilot, Steve Earley; Urban power outage: ©Iwan Baan/Reportage by Getty Images; Road washed out due to flooding: ©John Wark

pg. 39–Flooded subway: ©William Vantuono, Railway Age Magazine

pg. 42–Mountain stream: ©Dan Sherwood/Design Pics/Corbis

pg. 45–Hydroelectric plant: ©James Christensen/Foto Natura/Minden Pictures/Corbis; Wind turbines and cows: ©John Epperson/The Denver Post/Getty Images

pg. 46–Man inspecting wheat: ©iStockPhoto.com/small_frog

pg. 47–Farmer with corn: ©iStockPhoto.com/ValentinRussanov

pg. 48–Salmon fishing on Klamath River: ©David McLain/Aurora Photos

pg. 49–Wild rice harvesting: © Phil Schermeister/Corbis; Man and girl surveying water line: ©Mike Brubaker

pg. 50–Mt. Rainier, WA: ©Tim Fitzharris/Minden Pictures/Corbis

pg. 51–Person walking in forest: ©Michele Westmorland/Corbis

pg. 53–Alaska wildfire: ©Daryl Pederson/AlaskaStock/Corbis; Man inspecting tree: ©Melanie Stetson Freeman/The Christian Science Monitor/Getty Images; Dead trees in forest: ©Pete McBride/National Geographic Society

pg. 54–Development along Colorado's Front Range: ©Ted Wood Photography; Wildfire approaching housing development: ©Elmer Frederick Fischer/Corbis

pg. 56–Mussels: ©Doug Sokell/Visuals Unlimited/Corbis; Forest: ©Kevin R. Morris/Corbis; Polar bears: ©Jenny E. Ross/Corbis; Pika: ©iStockPhoto.com/Global_Exposure; Quaking aspen trees: ©Adam Jones/Visuals Unlimited/Corbis; Caribou calf: ©Matthias Breiter/Minden Pictures/Corbis; Hawaiian mountain vegetation: ©Michael Interisano/Design Pics/Corbis

pg. 57–Flying squirrel: ©Stephen Dalton/Minden Pictures/Corbis; Commercial fisher: ©Jeffrey Rotman/Corbis; Sunflowers: ©Annie Griffiths Belt/Corbis; Black rat snake: ©Gary Meszaros/Visuals Unlimited/Corbis; Mother bird and chick: ©Ronald Thompson/Frank Lane Picture Agency/Corbis; Two birds in water: ©Arthur Morris/Corbis; Tree seedling: ©Philip Gould/Corbis; Lionfish: ©Bruce Smith/AP/Corbis

pg. 59–Ocean: ©iStockPhoto.com/DigiClicks

pg. 60–Coral bleaching: courtesy of NOAA

pg. 61–Fishing vessel: ©iStockPhoto.com/mayo5

pg. 63–People discussing science findings: courtesy Armando Rodriguez, Miami-Dade County

pg. 65–Men installing solar panels: ©Don Mason/Blend Images/Corbis; Nuclear power plant: ©Joseph Sohm/Visions of America/Corbis; Wind turbines at sunset: ©Layne Kennedy/Corbis; Workers on automobile assembly line: ©Joseph Sohm/Visions of America/Corbis; Smog over city: ©iStockPhoto.com/SteinPhoto

pg. 66–Man assembling window: ©Carlos Osorio/AP/Corbis

pg. 68–Denver water system: ©Photo courtesy Denver Water

pg. 69–Ocean: ©iStockPhoto.com/AndrewJohnson

pg. 70–Autumn forest: ©Frank Siteman/Science Faction/Corbis

pg. 71–Stormwater wetland in Philadelphia: ©Louis Cook for PWD

pg. 72–Beach: ©Richard H. Cohen/Corbis

pg. 73–Clayton County, GA water recycling project: ©CCWA

pg. 74–Midwest farm: ©iStockPhoto.com/George_Burga

pg. 75–Flood in Cedar Falls: ©American Red Cross_Flickr

pg. 76–Bison in field: ©USFWS; Man and mailbox in flood: ©Lane Hickenbottom/Reuters/Corbis

pg. 77–Officer walking across cracked lakebed: ©Tony Gutierrez/AP/Corbis; Lakota tribe girl: ©Aaron Huey

pg. 78–Southwest image: ©Momatiuk-Eastcott/Corbis; Southwest image: ©Momatiuk-Eastcott/Corbis; Firefighters and wildfire: ©Frans Lanting/Corbis

pg. 80–Northwest image: Bryant Olsen, USFWS; Salmon: courtesy NOAA

pg. 81–Woman and oyster harvest: ©Macduff Everton/Corbis; Estuary restoration: Jesse Barham, U.S. Fish and Wildlife Service

pg. 82–Alaska image: ©Bryan F. Peterson/Corbis; Inupiaq seal hunter: ©Daniel Glick

pg. 83–Shore-protection structure: ©Carl Schoch; Newtok, Shishmaref village: ©Ned Rozell

pg. 84–Hawaiian image: ©Michael Wells/fstop/Corbis; Ko`olau Mountains, Oahu, HI: ©kstrebor via Flickr; Laysan Island, Papahānaumokuākea Marine National Monument: Andy Collins, NOAA

pg. 85–Coral reef: ©Ron Dahlquist/Corbis; Hawaiian waterfall: ©Air Maui

pg. 86–Aerial farm view: ©W. Perry Conway/Corbis

pg. 87–Flooded corn field: ©Nati Harnik/AP/Corbis; River flood waters: ©STR/Reuters/Corbis

pg. 88–Coastal image: ©Ocean/Corbis; Coastal road damage: ©John Tlumacki/The Boston Globe via Getty Images

pg. 89–Natural gas platform: ©Eric Kulin/First Light/Corbis

pg. 91–New York City coastal ecosystem restoration: ©Department of City Planning, New York City

pg. 94–Women discussing science findings: ©Lynn Laws Iowa State University 2013; Ongoing monitoring and observations: courtesy NOAA/NCDC

pg. 96–Men near culvert: ©Esperanza Stancioff, UMaine Extension and Maine Sea Grant; New York City bus: ©Najlah Feanny/Corbis

pg. 97– Women with rooftop garden: ©Denise Applewhite, Princeton Univ.; Southwest solar panels: ©Michael DeYoung/Blend Images/Corbis; Wind turbines: ©Jerome Levitch/Corbis

Back Cover–Field: ©Timothy Hearsum/AgStock Images/Corbis; Woman and solar panel: ©Bill Miles/Mint Images/Corbis; Sea ice melt: ©Steve Morgan/epa/Corbis; Flood rescue workers and victim: ©Adam Hunger/Reuters/Corbis

REFERENCES

The information presented in this *Highlights* report is derived directly from *Climate Change Impacts in the United States* - the full version of the Third National Climate Assessment. Thus the primary sources for *Highlights* are the relevant chapters of the full report and the external sources cited therein. In many cases, material selected for *Highlights* included direct references to external sources, and those references are provided below.

Overview

Numbered references for the Overview indicate the chapters from the full report that provide supporting evidence for the reported conclusions.

1. Ch. 2.

2. Ch. 2, 3, 6, 9, 20.

3. Ch. 2, 3, 4, 5, 6, 9, 10, 12, 16, 20, 24, 25.

4. Ch. 2, 12, 16, 18, 19, 20, 21, 22, 23.

5. Ch. 2, 4, 12, 16, 17, 18, 19, 20, 22, 25.

6. Ch. 2, 4, 5, 10, 12, 16, 17, 20, 22, 25.

7. Ch. 2, 12, 23, 24, 25.

8. Ch. 2, 12, 13, 14, 18, 19.

9. Ch. 2, 3, 12, 16, 17, 18, 19, 20, 21, 23.

10. Ch. 2, 9, 11, 12, 13, 16, 18, 19, 20, 25.

11. Ch. 3, 6, 8, 12, 14, 23, 24, 25.

12. Ch. 3, 7, 8, 25.

13. Ch. 2, 26, 27.

14. Ch. 26, 27, 28.

15. Ch. 2, 4, 27.

16. Ch. 2, 3, 5, 9, 11, 12, 13, 25, 26, 27, 28.

17. Ch. 3, 7, 9, 10, 12, 18, 20, 21, 26, 28.

18. Ch. 28.

19. Ch. 29, Appendix 6.

20. Ch. 30.

21. Ch. 2, Appendices 3 and 4.

22. Ch. 2, 16, 17, 18, 19, 20, 23, Appendices 3 and 4.

23. Ch. 2, 27, Appendices 3 and 4.

24. Ch. 3, 4, 5, 6, 7, 8, 9, 10, 11, 12, 13, 14, 16, 17, 18, 19, 20, 21, 22, 23, 24, 25.

25. Ch. 2, 6, 9, 11, 12, 16, 19, 20, 22, 23.

26. Ch. 2, 3, 5, 6, 11, 12, 16, 17, 18, 19, 20, 21, 22, 23, 25.

27. Ch. 2, 3, 12, 16, 17, 18, 19, 20, 21, 23.

28. Ch. 2, 6, 12, 13, 14, 18, 19.

29. Ch. 12, 17, 20, 21, 22, 23, 25.

30. Ch. 2, 3, 6, 7, 8, 10, 11, 14, 15, 19, 25.

31. Ch. 2, 12, 23, 24, 25.

32. Ch. 6, 7, 8, 9, 10, 13, 15, 25, 26, 27, 28.

Letter references refer to external sources

a. Kennedy, J. J., P. W. Thorne, T. C. Peterson, R. A. Reudy, P. A. Stott, D. E. Parker, S. A. Good, H. A. Titchner, and K. M Willett, 2010: How do we know the world has warmed? State of the Climate in 2009. *Bulletin of the American Meteorological Society*, **91**, S26-27, doi:10.1175/BAMS-91-7-StateoftheClimate. [Available online at http://journals.ametsoc.org/doi/abs/10.1175/BAMS-91-7-StateoftheClimate]

b. Huber, M., and R. Knutti, 2012: Anthropogenic and natural warming inferred from changes in Earth's energy balance. *Nature Geoscience*, **5**, 31-36, doi:10.1038/ngeo1327. [Available online at http://www.nature.com/ngeo/journal/v5/n1/pdf/ngeo1327.pdf]

c. Karl, T. R., J. T. Melillo, and T. C. Peterson, Eds., 2009: *Global Climate Change Impacts in the United States*. Cambridge University Press, 189 pp. [Available online at http://downloads.globalchange.gov/usimpacts/pdfs/climate-impacts-report.pdf]

d. Feely, R. A., S. C. Doney, and S. R. Cooley, 2009: Ocean acidification: Present conditions and future changes in a high-CO_2 world. *Oceanography*, **22**, 36-47, doi:10.5670/oceanog.2009.95. [Available online at http://www.tos.org/oceanography/archive/22-4_feely.pdf]

e. Bednaršek, N., G. A. Tarling, D. C. E. Bakker, S. Fielding, E. M. Jones, H. J. Venables, P. Ward, A. Kuzirian, B. Lézé, R. A. Feely, and E. J. Murphy, 2012: Extensive dissolution of live pteropods in the Southern Ocean. *Nature Geoscience*, **5**, 881-885, doi:10.1038/ngeo1635

Finding 1: Our Changing Climate

1. Karl, T. R., J. T. Melillo, and T. C. Peterson, Eds., 2009: *Global Climate Change Impacts in the United States*. Cambridge University Press, 189 pp. [Available online at http://downloads.globalchange.gov/usimpacts/pdfs/climate-impacts-report.pdf]

2. NSIDC, cited 2012: Arctic Sea Ice Reaches Lowest Extent for the Year and the Satellite Record. The National Snow and Ice Data Center. [Available online at http://nsidc.org/news/press/2012_seaiceminimum.html]

3. Kwok, R., and D. A. Rothrock, 2009: Decline in Arctic sea ice thickness from submarine and ICESat records: 1958–2008. *Geophysical Research Letters*, **36**, L15501, doi:10.1029/2009gl039035. [Available online at http://onlinelibrary.wiley.com/doi/10.1029/2009GL039035/pdf]

4. Maslanik, J., J. Stroeve, C. Fowler, and W. Emery, 2011: Distribution and trends in Arctic sea ice age through spring 2011. *Geophysical Research Letters*, **38**, L13502, doi:10.1029/2011gl047735. [Available online at http://onlinelibrary.wiley.com/doi/10.1029/2011GL047735/pdf]

5. Liu, J., J. A. Curry, H. Wang, M. Song, and R. M. Horton, 2012: Impact of declining Arctic sea ice on winter snowfall. *Proceedings of the National Academy of Sciences*, **109**, 4074-4079, doi:10.1073/pnas.1114910109. [Available online at http://www.pnas.org/content/109/11/4074.full.pdf+html]

6. Francis, J. A., and S. J. Vavrus, 2012: Evidence linking Arctic amplification to extreme weather in mid-latitudes. *Geophysical Research Letters*, **39**, L06801, doi:10.1029/2012GL051000. [Available online at http://onlinelibrary.wiley.com/doi/10.1029/2012GL051000/pdf]

7. Screen, J. A., and I. Simmonds, 2013: Exploring links between Arctic amplification and mid-latitude weather. *Geophysical Research Letters*, **40**, 959-964, doi:10.1002/grl.50174. [Available online at http://onlinelibrary.wiley.com/doi/10.1002/grl.50174/pdf]

8. NASA Earth Observatory, cited 2012: Visualizing the 2012 Sea Ice Minimum. NASA Earth Observatory, EOS Project Science Office, NASA Goddard Space Flight Center. [Available online at http://earthobservatory.nasa.gov/IOTD/view.php?id=79256]

9. Wouters, B., J. L. Bamber, M. R. van den Broeke, J. T. M. Lenaerts, and I. Sasgen, 2013: Limits in detecting acceleration of ice sheet mass loss due to climate variability. *Nature Geoscience*, **6**, 613-616, doi:10.1038/ngeo1874.

10. Peñuelas, J., T. Rutishauser, and I. Filella, 2009: Phenology feedbacks on climate change. *Science*, **324**, 887-888, doi:10.1126/science.1173004. [Available online at http://www.sciencemag.org/content/324/5929/887.short]

11. Ziska, L., K. Knowlton, C. Rogers, D. Dalan, N. Tierney, M. A. Elder, W. Filley, J. Shropshire, L. B. Ford, C. Hedberg, P. Fleetwood, K. T. Hovanky, T. Kavanaugh, G. Fulford, R. F. Vrtis, J. A. Patz, J. Portnoy, F. Coates, L. Bielory, and D. Frenz, 2011: Recent warming by latitude associated with increased length of ragweed pollen season in central North America. *Proceedings of the National Academy of Sciences*, **108**, 4248-4251, doi:10.1073/pnas.1014107108. [Available online at http://www.pnas.org/content/108/10/4248.full.pdf+html]

12. Hu, J. I. A., D. J. P. Moore, S. P. Burns, and R. K. Monson, 2010: Longer growing seasons lead to less carbon sequestration by a subalpine forest. *Global Change Biology*, **16**, 771-783, doi:10.1111/j.1365-2486.2009.01967.x. [Available online at http://onlinelibrary.wiley.com/doi/10.1111/j.1365-2486.2009.01967.x/pdf]

13. Frisvold, G., L. E. Jackson, J. G. Pritchett, and J. Ritten, 2013: Ch. 11: Agriculture and ranching. *Assessment of Climate Change in the Southwest United States: A Report Prepared for the National Climate Assessment*, G. Garfin, A. Jardine, R. Merideth, M. Black, and S. LeRoy, Eds., Island Press, 218-239. [Available online at http://swccar.org/sites/all/themes/files/SW-NCA-color-FINALweb.pdf]

14. Forster, P., V. Ramaswamy, P. Artaxo, T. Berntsen, R. Betts, D. W. Fahey, J. Haywood, J. Lean, D. C. Lowe, G. Myhre, J. Nganga, R. Prinn, G. Raga, M. Schulz, and R. Van Dorland, 2007: Ch. 2: Changes in atmospheric constituents and in radiative forcing. *Climate Change 2007: The Physical Science Basis. Contribution of Working Group I to the Fourth Assessment Report (AR4) of the Intergovernmental Panel on Climate Change*, S. Solomon, D. Qin, M. Manning, Z. Chen, M. Marquis, K. B. Averyt, M. Tignor, and H. L. Miller, Eds., Cambridge University Press. [Available online at http://www.ipcc.ch/publications_and_data/ar4/wg1/en/ch2.html]

15. Boden, T., G. Marland, and B. Andres, 2012: *Global CO₂ Emissions from Fossil-Fuel Burning, Cement Manufacutre, and Gas Flaring: 1751-2009*. Carbon Dioxide Information Analysis Center, Oak Ridge national Laboratory. [Available online at http://cdiac.ornl.gov/ftp/ndp030/global.1751_2009.ems]

Finding 2: Extreme Weather

1. Christidis, N., P. A. Stott, and S. J. Brown, 2011: The role of human activity in the recent warming of extremely warm daytime temperatures. *Journal of Climate*, **24**, 1922-1930, doi:10.1175/2011JCLI4150.1.

   Duffy, P. B., and C. Tebaldi, 2012: Increasing prevalence of extreme summer temperatures in the U.S. *Climatic Change*, **111**, 487-495, doi:10.1007/s10584-012-0396-6.

2. NCDC, cited 2012: Climate Data Online. National Climatic Data Center. [Available online at http://www.ncdc.noaa.gov/cdo-web/]

3. Hoerling, M., M. Chen, R. Dole, J. Eischeid, A. Kumar, J. W. Nielsen-Gammon, P. Pegion, J. Perlwitz, X.-W. Quan, and T. Zhang, 2013: Anatomy of an extreme event. *Journal of Climate*, 26, 2811–2832, doi:10.1175/JCLI-D-12-00270.1. [Available online at http://journals.ametsoc.org/doi/pdf/10.1175/JCLI-D-12-00270.1]

4. Sheffield, J., E. F. Wood, and M. L. Roderick, 2012: Little change in global drought over the past 60 years. *Nature*, **491**, 435-438, doi:10.1038/nature11575. [Available online at http://www.nature.com/nature/journal/v491/n7424/pdf/nature11575.pdf]

5. Mueller, B., and S. I. Seneviratne, 2012: Hot days induced by precipitation deficits at the global scale. *Proceedings of the National Academy of Sciences*, **109**, 12398-12403, doi:10.1073/pnas.1204330109. [Available online at http://www.pnas.org/content/109/31/12398.full.pdf+html]

6. Dai, A., 2006: Recent climatology, variability, and trends in global surface humidity. *Journal of Climate*, **19**, 3589-3606, doi:10.1175/JCLI3816.1. [Available online at http://journals.ametsoc.org/doi/pdf/10.1175/JCLI3816.1]

   Santer, B. D., C. Mears, F. J. Wentz, K. E. Taylor, P. J. Gleckler, T. M. L. Wigley, T. P. Barnett, J. S. Boyle, W. Brüggemann, N. P. Gillett, S. A. Klein, G. A. Meehl, T. Nozawa, D. W. Pierce, P. A. Stott, W. M. Washington, and M. F. Wehner, 2007: Identification of human-induced changes in atmospheric moisture content. *Proceedings of the National Academy of Sciences*, **104**, 15248-15253, doi:10.1073/pnas.0702872104. [Available online at http://sa.indiaenvironmentportal.org.in/files/file/PNAS-2007-Santer-15248-53.pdf]

   Simmons, A. J., K. M. Willett, P. D. Jones, P. W. Thorne, and D. P. Dee, 2010: Low-frequency variations in surface atmospheric humidity, temperature, and precipitation: Inferences from reanalyses and monthly gridded observational data sets. *Journal of Geophysical Research*, **115**, 1-21, doi:10.1029/2009JD012442.

   Willett, K. M., P. D. Jones, N. P. Gillett, and P. W. Thorne, 2008: Recent changes in surface humidity: Development of the HadCRUH dataset. *Journal of Climate*, **21**, 5364-5383, doi:10.1175/2008JCLI2274.1.

7. Kunkel, K. E., T.R. Karl, H. Brooks, J. Kossin, J. Lawrimore, D. Arndt, L. Bosart, D. Changnon, S.L. Cutter, N. Doesken, K. Emanuel, P.Ya. Groisman, R.W. Katz, T. Knutson, J. O'Brien, C. J. Paciorek, T. C. Peterson, K. Redmond, D. Robinson, J. Trapp, R. Vose, S. Weaver, M. Wehner, K. Wolter, and D. Wuebbles, 2013: Monitoring and understanding trends in extreme storms: State of knowledge. *Bulletin of the American Meteorological Society*, **94**, doi:10.1175/BAMS-D-11-00262.1. [Available online at http://journals.ametsoc.org/doi/pdf/10.1175/BAMS-D-11-00262.1]

8. NOAA, 2013: United States Flood Loss Report - Water Year 2011, 10 pp., National Oceanic and Atmospheric Administration, National Weather Service. [Available online at http://www.nws.noaa.gov/hic/summaries/WY2011.pdf]

9. Doocy, S., A. Daniels, S. Murray, and T. D. Kirsch, 2013: The human impact of floods: A historical review of events 1980-2009 and systematic literature review. *PLOS Currents Disasters*, doi:10.1371/currents.dis.f4deb457904936b07c09daa98ee8171a. [Available online at http://currents.plos.org/disasters/article/the-human-impact-of-floods-a-historical-review-of-events-1980-2009-and-systematic-literature-review/pdf]

10. Ashley, S. T., and W. S. Ashley, 2008: Flood fatalities in the United States. *Journal of Applied Meteorology and Climatology*, **47**, 805-818, doi:10.1175/2007JAMX1611.1. [Available online at http://journals.ametsoc.org/doi/pdf/10.1175/2007JAMC1611.1]

11. Hirsch, R. M., and K. R. Ryberg, 2012: Has the magnitude of floods across the USA changed with global CO₂ levels? *Hydrological Sciences Journal*, **57**, 1-9, doi:10.1080/02626667.2011.621895. [Available online at http://www.tandfonline.com/doi/abs/10.1080/02626667.2011.621895]

U.S. GLOBAL CHANGE RESEARCH PROGRAM

Villarini, G., F. Serinaldi, J. A. Smith, and W. F. Krajewski, 2009: On the stationarity of annual flood peaks in the continental United States during the 20th century. *Water Resources Research*, **45**, W08417, doi:10.1029/2008wr007645. [Available online at http://onlinelibrary.wiley.com/doi/10.1029/2008WR007645/pdf]

Villarini, G., and J. A. Smith, 2010: Flood peak distributions for the eastern United States. *Water Resources Research*, **46**, W06504, doi:10.1029/2009wr008395. [Available online at http://onlinelibrary.wiley.com/doi/10.1029/2009WR008395/pdf]

Villarini, G., J. A. Smith, M. L. Baeck, and W. F. Krajewski, 2011: Examining flood frequency distributions in the Midwest U.S. *JAWRA Journal of the American Water Resources Association*, **47**, 447-463, doi:10.1111/j.1752-1688.2011.00540.x. [Available online at http://onlinelibrary.wiley.com/doi/10.1111/j.1752-1688.2011.00540.x/pdf]

12. Peterson, T. C., R. R. Heim, R. Hirsch, D. P. Kaiser, H. Brooks, N. S. Diffenbaugh, R. M. Dole, J. P. Giovannettone, K. Guirguis, T. R. Karl, R. W. Katz, K. Kunkel, D. Lettenmaier, G. J. McCabe, C. J. Paciorek, K. R. Ryberg, S. Schubert, V. B. S. Silva, B. C. Stewart, A. V. Vecchia, G. Villarini, R. S. Vose, J. Walsh, M. Wehner, D. Wolock, K. Wolter, C. A. Woodhouse, and D. Wuebbles, 2013: Monitoring and understanding changes in heat waves, cold waves, floods and droughts in the United States: State of knowledge. *Bulletin American Meteorological Society*, **94**, 821-834, doi:10.1175/BAMS-D-12-00066.1. [Available online at http://journals.ametsoc.org/doi/pdf/10.1175/BAMS-D-12-00066.1]

13. Bell, G. D., E. S. Blake, C. W. Landsea, T. B. Kimberlain, S. B. Goldenberg, J. Schemm, and R. J. Pasch, 2012: [Tropical cyclones] Atlantic basin [in "State of the Climate in 2011"]. *Bulletin of the American Meteorological Society*, **93**, S99-S105, doi:10.1175/2012BAMSStateoftheClimate.1. [Available online at http://www1.ncdc.noaa.gov/pub/data/cmb/bams-sotc/climate-assessment-2011-lo-rez.pdf]

Landsea, C. W., and J. L. Franklin, 2013: Atlantic hurricane database uncertainty and presentation of a new database format. *Monthly Weather Review*, **141**, 3576-3592, doi:10.1175/MWR-D-12-00254.1. [Available online at http://journals.ametsoc.org.prox.lib.ncsu.edu/doi/pdf/10.1175/MWR-D-12-00254.1]

Torn, R. D., and C. Snyder, 2012: Uncertainty of tropical cyclone best-track information. *Weather and Forecasting*, **27**, 715-729, doi:10.1175/waf-d-11-00085.1. [Available online at http://journals.ametsoc.org/doi/pdf/10.1175/WAF-D-11-00085.1]

14. Emanuel, K., and A. Sobel, 2013: Response of tropical sea surface temperature, precipitation, and tropical cyclone-related variables to changes in global and local forcing. *Journal of Advances in Modeling Earth Systems*, **5**, 447-458, doi:10.1002/jame.20032. [Available online at http://onlinelibrary.wiley.com/doi/10.1002/jame.20032/pdf]

Zhang, R., and T. L. Delworth, 2009: A new method for attributing climate variations over the Atlantic Hurricane Basin's main development region. *Geophysical Research Letters*, **36**, 5, doi:10.1029/2009GL037260.

15. Camargo, S. J., M. Ting, and Y. Kushnir, 2013: Influence of local and remote SST on North Atlantic tropical cyclone potential intensity. *Climate Dynamics*, **40**, 1515-1529, doi:10.1007/s00382-012-1536-4.

Ramsay, H. A., and A. H. Sobel, 2011: Effects of relative and absolute sea surface temperature on tropical cyclone potential intensity using a single-column model. *Journal of Climate*, **24**, 183-

193, doi:10.1175/2010jcli3690.1. [Available online at http://journals.ametsoc.org/doi/pdf/10.1175/2010JCLI3690.1]

Vecchi, G. A., A. Clement, and B. J. Soden, 2008: Examining the tropical Pacific's response to global warming. *Eos, Transactions, American Geophysical Union*, **89**, 81-83, doi:10.1029/2008EO090002.

Vecchi, G. A., and B. J. Soden, 2007: Effect of remote sea surface temperature change on tropical cyclone potential intensity. *Nature*, **450**, 1066-1070, doi:10.1038/nature06423.

16. Vose, R. S., S. Applequist, M. A. Bourassa, S. C. Pryor, R. J. Barthelmie, B. Blanton, P. D. Bromirski, H. E. Brooks, A. T. DeGaetano, R. M. Dole, D. R. Easterling, R. E. Jensen, T. R. Karl, R. W. Katz, K. Klink, M. C. Kruk, K. E. Kunkel, M. C. MacCracken, T. C. Peterson, K. Shein, B. R. Thomas, J. E. Walsh, X. L. Wang, M. F. Wehner, D. J. Wuebbles, and R. S. Young, 2013: Monitoring and understanding changes in extremes: Extratropical storms, winds, and waves. *Bulletin of the American Meteorological Society*, **in press**, doi:10.1175/BAMS-D-12-00162.1. [Available online at http://journals.ametsoc.org/doi/pdf/10.1175/BAMS-D-12-00162.1]

17. Wang, X. L., Y. Feng, G. P. Compo, V. R. Swail, F. W. Zwiers, R. J. Allan, and P. D. Sardeshmukh, 2012: Trends and low frequency variability of extra-tropical cyclone activity in the ensemble of twentieth century reanalysis. *Climate Dynamics*, 1-26, doi:10.1007/s00382-012-1450-9.

Wang, X. L., V. R. Swail, and F. W. Zwiers, 2006: Climatology and changes of extratropical cyclone activity: Comparison of ERA-40 with NCEP-NCAR reanalysis for 1958-2001. *Journal of Climate*, **19**, 3145-3166, doi:10.1175/JCLI3781.1. [Available online at http://journals.ametsoc.org/doi/abs/10.1175/JCLI3781.1]

18. NOAA, cited 2013: Billion Dollar Weather/Climate Disasters. National Oceanic and Atmospheric Administration [Available online at http://www.ncdc.noaa.gov/billions]

Finding 3: Future Climate

1. Dai, A., 2012: Increasing drought under global warming in observations and models. *Nature Climate Change*, **3**, 52-58, doi:10.1038/nclimate1633. [Available online at http://www.nature.com/nclimate/journal/vaop/ncurrent/full/nclimate1633.html?utm_source=feedblitz&utm_medium=FeedBlitzEmail&utm_content=559845&utm_campaign=0]

Liang, X., D. P. Lettenmaier, E. F. Wood, and S. J. Burges, 1994: A simple hydrologically based model of land surface water and energy fluxes for general circulation models. *Journal of Geophysical Research*, **99**, 14415-14428, doi:10.1029/94JD00483. [Available online at http://onlinelibrary.wiley.com/doi/10.1029/94JD00483/pdf]

Liang, X., E. F. Wood, and D. P. Lettenmaier, 1996: Surface soil moisture parameterization of the VIC-2L model: Evaluation and modification. *Global and Planetary Change*, **13**, 195-206, doi:10.1016/0921-8181(95)00046-1.

Maurer, E. P., A. W. Wood, J. C. Adam, D. P. Lettenmaier, and B. Nijssen, 2002: A long-term hydrologically based dataset of land surface fluxes and states for the conterminous United States. *Journal of Climate*, **15**, 3237-3251, doi:10.1175/1520-0442(2002)015<3237:ALTHBD>2.0.CO;2. [Available online at http://journals.ametsoc.org/doi/pdf/10.1175/1520-0442(2002)015%3C3237%3AALTHBD%3E2.0.CO%3B2]

Nijssen, B., D. P. Lettenmaier, X. Liang, S. W. Wetzel, and E. F. Wood, 1997: Streamflow simulation for continental-scale river basins. *Water Resources Research*, **33**, 711-724, doi:10.1029/96WR03517. [Available online at http://www.agu.org/pubs/crossref/1997/96WR03517. shtml]

Wood, A. W., A. Kumar, and D. P. Lettenmaier, 2005: A retrospective assessment of National Centers for Environmental Prediction climate model–based ensemble hydrologic forecasting in the western United States. *Journal of Geophysical Research*, **110**, 16, doi:10.1029/2004JD004508.

Wood, A. W., and D. P. Lettenmaier, 2006: A test bed for new seasonal hydrologic forecasting approaches in the western United States. *Bulletin of the American Meteorological Society*, **87**, 1699-1712, doi:10.1175/BAMS-87-12-1699.

2. Church, J. A., N. J. White, L. F. Konikow, C. M. Domingues, J. G. Cogley, E. Rignot, J. M. Gregory, M. R. van den Broeke, A. J. Monaghan, and I. Velicogna, 2011: Revisiting the Earth's sea-level and energy budgets from 1961 to 2008. *Geophysical Research Letters*, **38**, L18601, doi:10.1029/2011GL048794.

3. AMAP, 2011: Snow, Water, Ice and Permafrost in the Arctic (SWIPA): Climate Change and the Cryosphere, 538 pp., Arctic Monitoring and Assessment Programme, Oslo, Norway. [Available online at http://www.amap.no/documents/download/968]

4. Kemp, A. C., B. P. Horton, J. P. Donnelly, M. E. Mann, M. Vermeer, and S. Rahmstorf, 2011: Climate related sea-level variations over the past two millennia. *Proceedings of the National Academy of Sciences*, **108**, 11017-11022, doi:10.1073/pnas.1015619108. [Available online at http://www.pnas.org/content/108/27/11017.full.pdf+html]

5. Church, J. A., and N. J. White, 2011: Sea-level rise from the late 19th to the early 21st century. *Surveys in Geophysics*, **32**, 585-602, doi:10.1007/s10712-011-9119-1.

6. Nerem, R. S., D. P. Chambers, C. Choe, and G. T. Mitchum, 2010: Estimating mean sea level change from the TOPEX and Jason altimeter missions. *Marine Geodesy*, **33**, 435-446, doi:10.1080/01490 419.2010.491031. [Available online at http://www.tandfonline.com/ doi/pdf/10.1080/01490419.2010.491031]

7. Parris, A., P. Bromirski, V. Burkett, D. Cayan, M. Culver, J. Hall, R. Horton, K. Knuuti, R. Moss, J. Obeysekera, A. Sallenger, and J. Weiss, 2012: Global Sea Level Rise Scenarios for the United States National Climate Assessment. NOAA Tech Memo OAR CPO-1, 37 pp., National Oceanic and Atmospheric Administration, Silver Spring, MD. [Available online at http://scenarios.globalchange.gov/ sites/default/files/NOAA_SLR_r3_0.pdf]

8. IPCC, 2007: *Climate Change 2007: The Physical Science Basis. Contribution of Working Group I to the Fourth Assessment Report of the Intergovernmental Panel on Climate Change.* S. Solomon, D. Qin, M. Manning, Z. Chen, M. Marquis, K. B. Averyt, M. Tignor, and H. L. Miller, Eds. Cambridge University Press, 996 pp. [Available online at http:// www.ipcc.ch/publications_and_data/publications_ipcc_fourth_ assessment_report_wg1_report_the_physical_science_basis.htm]

Collins, M., R. Knutti, J. M. Arblaster, J.-L. Dufresne, T. Fichefet, F. P., X. Gao, W. J. Gutowski, T. Johns, G. Krinner, M. Shongwe, C. Tebaldi, A. J. Weaver, and M. Wehner, 2013: Long-term climate change: Projections, commitments and irreversibility. *Climate Change 2013: The Physical Science Basis. Contribution of Working Group I to the Fifth Assessment Report of the Intergovernmental Panel on Climate Change*, T. F. Stocker, D. Qin, G.-K. Plattner, M. Tignor, S. K. Allen, J. Boschung, Nauels, Y. Xia, V. Bex, and P. M. Midgley, Eds., Cambridge University Press. [Available online at http://www. climatechange2013.org/report/review-drafts/]

Finding 4: Widespread Impacts

1. Lister, S. A., 2005: Hurricane Katrina: The Public Health and Medical Response, 24 pp., Congressional Research Service Report for Congress. [Available online at http://fpc.state.gov/documents/ organization/54255.pdf]

2. Anderson, G. B., and M. L. Bell, 2012: Lights out: Impact of the August 2003 power outage on mortality in New York, NY. *Epidemiology*, **23**, 189-193, doi:10.1097/EDE.0b013e318245c61c.

3. Ostro, B., S. Rauch, R. Green, B. Malig, and R. Basu, 2010: The effects of temperature and use of air conditioning on hospitalizations. *American Journal of Epidemiology*, **172**, 1053-1061, doi:10.1093/ aje/kwq231. [Available online at http://aje.oxfordjournals.org/ content/172/9/1053.full.pdf+html]

4. Hess, J. J., J. Z. McDowell, and G. Luber, 2012: Integrating climate change adaptation into public health practice: Using adaptive management to increase adaptive capacity and build resilience. *Environmental Health Perspectives*, **120**, 171-179, doi:10.1289/ ehp.1103515. [Available online at http://www.ncbi.nlm.nih.gov/ pmc/articles/PMC3279431/]

5. Shonkoff, S. B., R. Morello-Frosch, M. Pastor, and J. Sadd, 2011: The climate gap: Environmental health and equity implications of climate change and mitigation policies in California—a review of the literature. *Climatic Change*, 485-503, doi:10.1007/s10584-011- 0310-7.

6. Kent, J. D., 2006: Louisiana Hurricane Impact Atlas, 39 pp., Louisiana Geographic Information Center, Baton Rouge, LA. [Available online at http://lagic.lsu.edu/lgisc/publications/2005/ LGISC-PUB-20051116-00_2005_HURRICANE_ATLAS.pdf]

7. Burke, L., L. Reytar, M. Spalding, and A. Perry, 2011: *Reefs at Risk Revisited*. World Resources Institute, 130 pp. [Available online at http://pdf.wri.org/reefs_at_risk_revisited.pdf]

Dudgeon, S. R., R. B. Aronson, J. F. Bruno, and W. F. Precht, 2010: Phase shifts and stable states on coral reefs. *Marine Ecology Progress Series*, **413**, 201-216, doi:10.3354/meps08751. [Available online at http://johnfbruno.web.unc.edu/files/2011/11/Dudgeon-et-al-MEPS-ASS-2010.pdf]

Frieler, K., M. Meinshausen, A. Golly, M. Mengel, K. Lebek, S. D. Donner, and O. Hoegh-Guldberg, 2013: Limiting global warming to 2°C is unlikely to save most coral reefs. *Nature Climate Change*, **3**, 165-170, doi:10.1038/nclimate1674.

Hoegh-Guldberg, O., P. J. Mumby, A. J. Hooten, R. S. Steneck, P. Greenfield, E. Gomez, C. D. Harvell, P. F. Sale, A. J. Edwards, K. Caldeira, N. Knowlton, C. M. Eakin, R. Iglesias-Prieto, N. Muthiga, R. H. Bradbury, A. Dubi, and M. E. Hatziolos, 2007: Coral reefs under rapid climate change and ocean acidification. *Science*, **318**, 1737-1742, doi:10.1126/science.1152509.

Hughes, T. P., N. A. J. Graham, J. B. C. Jackson, P. J. Mumby, and R. S. Steneck, 2010: Rising to the challenge of sustaining coral reef resilience. *Trends in Ecology & Evolution*, **25**, 633-642, doi:10.1016/j. tree.2010.07.011.

8. Mumby, P. J., and R. S. Steneck, 2011: The resilience of coral reefs and its implications for reef management. *Coral Reefs: An Ecosystem in Transition*, Z. Dubinsky, and N. Stambler, Eds., 509-519.

9. Gardner, T. A., I. M. Côté, J. A. Gill, A. Grant, and A. R. Watkinson, 2003: Long-term region-wide declines in Caribbean corals. *Science*, **301,** 958-960, doi:10.1126/science.1086050.

10. Alvarez-Filip, L., N. K. Dulvy, J. A. Gill, I. M. Côté, and A. R. Watkinson, 2009: Flattening of Caribbean coral reefs: Region-wide declines in architectural complexity. *Proceedings of the Royal Society B: Biological Sciences*, **276,** 3019-3025, doi:10.1098/rspb.2009.0339. [Available online at http://rspb.royalsocietypublishing.org/content/276/1669/3019.full.pdf+html]

11. Miller, J., E. Muller, C. Rogers, R. Waara, A. Atkinson, K. R. T. Whelan, M. Patterson, and B. Witcher, 2009: Coral disease following massive bleaching in 2005 causes 60% decline in coral cover on reefs in the US Virgin Islands. *Coral Reefs*, **28,** 925-937, doi:10.1007/s00338-009-0531-7. [Available online at http://link.springer.com/content/pdf/10.1007%2Fs00338-009-0531-7]

Weil, E., A. Croquer, and I. Urreiztieta, 2009: Temporal variability and impact of coral diseases and bleaching in La Parguera, Puerto Rico from 2003–2007. *Caribbean Journal of Science*, **45,** 221-246. [Available online at http://caribjsci.org/45_2_3/45_221-246.pdf]

12. Sandin, S. A., J. E. Smith, E. E. DeMartini, E. A. Dinsdale, S. D. Donner, A. M. Friedlander, T. Konotchick, M. Malay, J. E. Maragos, D. Obura, O. Pantos, G. Paulay, M. Richie, F. Rohwer, R. E. Schroeder, S. Walsh, J. B. C. Jackson, N. Knowlton, and E. Sala, 2008: Baselines and degradation of coral reefs in the northern Line Islands. *PLoS ONE*, **3,** 1-11, doi:10.1371/journal.pone.0001548. [Available online at http://www.pifsc.noaa.gov/library/pubs/Sandin_etal_PLosONE_2008.pdf]

13. ERCOT, 2011: Grid Operations and Planning Report (Austin, Texas, December 12-13, 2011), 25 pp., Electric Reliability Council of Texas. [Available online at http://www.ercot.com/content/meetings/board/keydocs/2011/1212/Item_06e_-_Grid_Operations_and_Planning_Report.pdf]

14. Giberson, M., cited 2012: Power Consumption Reaches New Peaks in Texas, ERCOT Narrowly Avoids Rolling Blackouts. The Energy Collective. [Available online at http://theenergycollective.com/michaelgiberson/63173/power-consumption-reaches-new-peaks-texas-ercot-narrowly-avoids-rolling-blacko]

Finding 5: Human Health

1. Dennekamp, M., and M. Carey, 2010: Air quality and chronic disease: Why action on climate change is also good for health. *New South Wales Public Health Bulletin*, **21,** 115-121, doi:10.1071/NB10026. [Available online at http://www.publish.csiro.au/?act=view_file&file_id=NB10026.pdf]

Kampa, M., and E. Castanas, 2008: Human health effects of air pollution. *Environmental Pollution*, **151,** 362-367, doi:10.1016/j.envpol.2007.06.012.

Kinney, P. L., 2008: Climate change, air quality and human health. *American Journal of Preventive Medicine*, **35,** 459-467, doi:10.1016/j.amepre.2008.08.025. [Available online at http://www.ajpmonline.org/article/S0749-3797%2808%2900690-9/fulltext]

Anderson, G. B., J. R. Krall, R. D. Peng, and M. L. Bell, 2012: Is the relation between ozone and mortality confounded by chemical components of particulate matter? Analysis of 7 components in 57 US communities. *American Journal of Epidemiology*, **176,** 726-732, doi:10.1093/aje/kws188. [Available online at http://aje.oxfordjournals.org/content/176/8/726.full.pdf+html]

2. Fiore, A. M., V. Naik, D. V. Spracklen, A. Steiner, N. Unger, M. Prather, D. Bergmann, P. J. Cameron-Smith, I. Cionni, W. J. Collins, S. Dalsoren, V. Eyring, G. A. Folberth, P. Ginoux, L. W. Horowitz, B. Josse, J.-F. Lamarque, I. A. MacKenzie, T. Nagashima, F. M. O'Connor, M. Righi, S. T. Rumbold, D. T. Shindell, R. B. Skeie, K. Sudo, S. Szopa, T. Takemura, and G. Zeng, 2012: Global air quality and climate. *Chemical Society Reviews*, **41,** 6663-6683, doi:10.1039/c2cs35095e.

Peel, J. L., R. Haeuber, V. Garcia, L. Neas, and A. G. Russell, 2012: Impact of nitrogen and climate change interactions on ambient air pollution and human health. *Biogeochemistry*, doi:10.1007/s10533-012-9782-4. [Available online at http://link.springer.com/content/pdf/10.1007%2Fs10533-012-9782-4]

3. NCDC, cited 2012: State of the Climate Wildfires. NOAA's National Climatic Data Center. [Available online at http://www.ncdc.noaa.gov/sotc/fire/2012/11]

4. Johnston, F. H., S. B. Henderson, Y. Chen, J. T. Randerson, M. Marlier, R. S. DeFries, P. Kinney, D. M. J. S. Bowman, and M. Brauer, 2012: Estimated global mortality attributable to smoke from landscape fires. *Environmental Health Perspectives*, **120,** 695-701, doi:10.1289/ehp.1104422. [Available online at http://www.ncbl.nlm.nih.gov/pmc/articles/PMC3346787/]

5. Littell, J. S., D. McKenzie, D. L. Peterson, and A. L. Westerling, 2009: Climate and wildfire area burned in western US ecoprovinces, 1916-2003. *Ecological Applications*, **19,** 1003-1021, doi:10.1890/07-1183.1.

MacDonald, G. M., 2010: Water, climate change, and sustainability in the southwest. *Proceedings of the National Academy of Sciences*, **107,** 21256-21262, doi:10.1073/pnas.0909651107. [Available online at http://www.pnas.org/content/107/50/21256.full.pdf]

Mills, D. M., 2009: Climate change, extreme weather events, and US health impacts: What can we say? *Journal of Occupational and Environmental Medicine*, **51,** 26-32, doi:10.1097/JOM.0b013e31817d32da.

Westerling, A. L., H. G. Hidalgo, D. R. Cayan, and T. W. Swetnam, 2006: Warming and earlier spring increase western U.S. forest wildfire activity. *Science*, **313,** 940-943, doi:10.1126/science.1128834.

Westerling, A. L., M. G. Turner, E. A. H. Smithwick, W. H. Romme, and M. G. Ryan, 2011: Continued warming could transform Greater Yellowstone fire regimes by mid-21st century. *Proceedings of the National Academy of Sciences*, **108,** 13165-13170, doi:10.1073/pnas.1110199108. [Available online at http://www.pnas.org/content/early/2011/07/20/1110199108.abstract; http://www.pnas.org/content/108/32/13165.full.pdf]

6. Shea, K. M., R. T. Truckner, R. W. Weber, and D. B. Peden, 2008: Climate change and allergic disease. *Journal of Allergy and Clinical Immunology*, **122,** 443-453, doi:10.1016/j.jaci.2008.06.032.

7. Trenberth, K. E., 2011: Changes in precipitation with climate change. *Climate Research*, **47,** 123-138, doi:10.3354/cr00953.

REFERENCES

8. Dennekamp, M., and M. J. Abramson, 2011: The effects of bushfire smoke on respiratory health. *Respirology*, **16,** 198-209, doi:10.1111/j.1440-1843.2010.01868.x.

9. Jaffe, D., D. Chand, W. Hafner, A. Westerling, and D. Spracklen, 2008: Influence of fires on O_3 concentrations in the western US. *Environmental Science & Technology*, **42,** 5885-5891, doi:10.1021/es800084k.

Jaffe, D., W. Hafner, D. Chand, A. Westerling, and D. Spracklen, 2008: Interannual variations in PM2.5 due to wildfires in the western United States. *Environmental Science & Technology*, **42,** 2812-2818, doi:10.1021/es702755v.

Pfister, G. G., C. Wiedinmyer, and L. K. Emmons, 2008: Impacts of the fall 2007 California wildfires on surface ozone: Integrating local observations with global model simulations. *Geophysical Research Letters*, **35,** L19814, doi:10.1029/2008GL034747.

Spracklen, D. V., J. A. Logan, L. J. Mickley, R. J. Park, R. Yevich, A. L. Westerling, and D. A. Jaffe, 2007: Wildfires drive interannual variability of organic carbon aerosol in the western US in summer. *Geophysical Research Letters*, **34,** L16816, doi:10.1029/2007GL030037.

10. Delfino, R. J., S. Brummel, J. Wu, H. Stern, B. Ostro, M. Lipsett, A. Winer, D. H. Street, L. Zhang, T. Tjoa, and D. L. Gillen, 2009: The relationship of respiratory and cardiovascular hospital admissions to the southern California wildfires of 2003. *Occupational and Environmental Medicine*, **66,** 189-197, doi:10.1136/oem.2008.041376. [Available online at [http://oem.bmj.com/content/66/3/189.full.pdf+html]

11. Elliott, C., S. Henderson, and V. Wan, 2013: Time series analysis of fine particulate matter and asthma reliever dispensations in populations affected by forest fires. *Environmental Health*, **12,** 11, doi:10.1186/1476-069X-12-11. [Available online at http://www.ehjournal.net/content/12/1/11]

12. Jenkins, J. L., E. B. Hsu, L. M. Sauer, Y. H. Hsieh, and T. D. Kirsch, 2009: Prevalence of unmet health care needs and description of health care-seeking behavior among displaced people after the 2007 California wildfires. *Disaster Medicine and Public Health Preparedness*, **3,** S24-28, doi:10.1097/DMP.0b013e31819f1afc. [Available online at http://www.dmphp.org/cgi/content/full/3/Supplement_1/S24]

Lee, T. S., K. Falter, P. Meyer, J. Mott, and C. Gwynn, 2009: Risk factors associated with clinic visits during the 1999 forest fires near the Hoopa Valley Indian Reservation, California, USA. *International Journal of Environmental Health Research*, **19,** 315-327, doi:10.1080/09603120802712750.

13. Henderson, S. B., M. Brauer, Y. C. Macnab, and S. M. Kennedy, 2011: Three measures of forest fire smoke exposure and their associations with respiratory and cardiovascular health outcomes in a population-based cohort. *Environmental Health Perspectives*, **119,** 1266-1271, doi:10.1289/ehp.1002288. [Available online at http://europepmc.org/articles/PMC3230386?pdf=render]

Holstius, D. M., C. E. Reid, B. M. Jesdale, and R. Morello-Frosch, 2012: Birth weight following pregnancy during the 2003 southern California wildfires. *Environmental Health Perspectives*, **120,** 1340-1345, doi:10.1289/ehp.110451. [Available online at http://www.ncbi.nlm.nih.gov/pmc/articles/PMC3440113/pdf/ehp.1104515.pdf]

Jacob, D. J., and D. A. Winner, 2009: Effect of climate change on air quality. *Atmospheric Environment*, **43,** 51-63, doi:10.1016/j.atmosenv.2008.09.051. [Available online at http://www.sciencedirect.com/science/article/pii/S1352231008008571]

Marlier, M. E., R. S. DeFries, A. Voulgarakis, P. L. Kinney, J. T. Randerson, D. T. Shindell, Y. Chen, and G. Faluvegi, 2013: El Nino and health risks from landscape fire emissions in southeast Asia. *Nature Climate Change*, **3,** 131-136, doi:10.1038/nclimate1658.

Rappold, A., W. Cascio, V. Kilaru, S. Stone, L. Neas, R. Devlin, and D. Diaz-Sanchez, 2012: Cardio-respiratory outcomes associated with exposure to wildfire smoke are modified by measures of community health. *Environmental Health*, **11,** 71, doi:10.1186/1476-069X-11-71. [Available online at http://www.ehjournal.net/content/pdf/1476-069X-11-71.pdf]

Westerling, A. L., and B. P. Bryant, 2008: Climate change and wildfire in California. *Climatic Change*, **87,** 231-249, doi:10.1007/s10584-007-9363-z.

14. Ziska, L., K. Knowlton, C. Rogers, D. Dalan, N. Tierney, M. A. Elder, W. Filley, J. Shropshire, L. B. Ford, C. Hedberg, P. Fleetwood, K. T. Hovanky, T. Kavanaugh, G. Fulford, R. F. Vrtis, J. A. Patz, J. Portnoy, F. Coates, L. Bielory, and D. Frenz, 2011: Recent warming by latitude associated with increased length of ragweed pollen season in central North America. *Proceedings of the National Academy of Sciences*, **108,** 4248-4251, doi:10.1073/pnas.1014107108. [Available online at http://www.pnas.org/content/108/10/4248.full.pdf+html]

15. Emberlin, J., M. Detandt, R. Gehrig, S. Jaeger, N. Nolard, and A. Rantio-Lehtimäki, 2002: Responses in the start of Betula (birch) pollen seasons to recent changes in spring temperatures across Europe. *International Journal of Biometeorology*, **46,** 159-170, doi:10.1007/s00484-002-0139-x.

Pinkerton, K. E., W. N. Rom, M. Akpinar-Elci, J. R. Balmes, H. Bayram, O. Brandli, J. W. Hollingsworth, P. L. Kinney, H. G. Margolis, W. J. Martin, E. N. Sasser, K. R. Smith, and T. K. Takaro, 2012: An official American Thoracic Society workshop report: Climate change and human health. *Proceedings of the American Thoracic Society*, **9,** 3-8, doi:10.1513/pats.201201-015ST. [Available online at http://www.atsjournals.org/doi/pdf/10.1513/pats.201201-015ST]

Schmier, J. K., and K. L. Ebi, 2009: The impact of climate change and aeroallergens on children's health. *Allergy and Asthma Proceedings*, 229-237 pp.

Sheffield, P. E., J. L. Carr, P. L. Kinney, and K. Knowlton, 2011: Modeling of regional climate change effects on ground-level ozone and childhood asthma. *American Journal of Preventive Medicine*, **41,** 251-257, doi:10.1016/j.amepre.2011.04.017. [Available online at http://download.journals.elsevierhealth.com/pdfs/journals/0749-3797/PIIS0749379711003461.pdf]

Sheffield, P. E., and P. J. Landrigan, 2011: Global climate change and children's health: Threats and strategies for prevention. *Environmental Health Perspectives*, **119,** 291-298, doi:10.1289/ehp.1002233. [Available online at http://environmentportal.in/files/climate%20change%20and%20childrens%20health.pdf]

16. Ariano, R., G. W. Canonica, and G. Passalacqua, 2010: Possible role of climate changes in variations in pollen seasons and allergic sensitizations during 27 years. *Annals of Allergy, Asthma & Immunology*, **104,** 215-222, doi:10.1016/j.anai.2009.12.005.

Breton, M. C., M. Garneau, I. Fortier, F. Guay, and J. Louis, 2006: Relationship between climate, pollen concentrations of *Ambrosia* and medical consultations for allergic rhinitis in Montreal, 1994–2002. *Science of The Total Environment*, **370**, 39-50, doi:10.1016/j.scitotenv.2006.05.022.

EPA, 2008: Review of the Impact of Climate Variability and Change on Aeroallergens and Their Associated Effects. EPA/600/R-06/164F, 125 pp., U.S. Environmental Protection Agency, Washington, D.C. [Available online at http://ofmpub.epa.gov/eims/eimscomm.getfile?p_download_id=490474]

17. Sheffield, P. E., K. R. Weinberger, K. Ito, T. D. Matte, R. W. Mathes, G. S. Robinson, and P. L. Kinney, 2011: The association of tree pollen concentration peaks and allergy medication sales in New York City: 2003–2008. *ISRN Allergy*, **2011**, 1-7, doi:10.5402/2011/537194. [Available online at http://downloads.hindawi.com/isrn/allergy/2011/537194.pdf]

18. Staudt, A., P. Glick, D. Mizejewski, and D. Inkley, 2010: Extreme Allergies and Global Warming, 12 pp., National Wildlife Federation and Asthma and Allergy Foundation of America. [Available online at http://www.nwf.org/~/media/PDFs/Global-Warming/Reports/NWF_AllergiesFinal.ashx]

19. D'amato, G., and L. Cecchi, 2008: Effects of climate change on environmental factors in respiratory allergic diseases. *Clinical & Experimental Allergy*, **38**, 1264-1274, doi:10.1111/j.1365-2222.2008.03033.x.

D'amato, G., L. Cecchi, M. D'amato, and G. Liccardi, 2010: Urban air pollution and climate change as environmental risk factors of respiratory allergy: An update. *Journal of Investigational Allergology and Clinical Immunology*, **20**, 95-102.

Reid, C. E., and J. L. Gamble, 2009: Aeroallergens, allergic disease, and climate change: Impacts and adaptation. *EcoHealth*, **6**, 458-470, doi:10.1007/s10393-009-0261-x. [Available online at http://link.springer.com/content/pdf/10.1007%2Fs10393-009-0261-x]

Nordling, E., N. Berglind, E. Melén, G. Emenius, J. Hallberg, F. Nyberg, G. Pershagen, M. Svartengren, M. Wickman, and T. Bellander, 2008: Traffic-related air pollution and childhood respiratory symptoms, function and allergies. *Epidemiology*, **19**, 401-408, doi:10.1097/EDE.0b013e31816a1ce3. [Available online at http://journals.lww.com/epidem/Fulltext/2008/05000/Traffic_Related_Air_Pollution_and_Childhood.11.aspx]

20. Fisk, W. J., Q. Lei-Gomez, and M. J. Mendell, 2007: Meta-analyses of the associations of respiratory health effects with dampness and mold in homes. *Indoor Air*, **17**, 284-296, doi:10.1111/j.1600-0668.2007.00475.x. [Available online at http://onlinelibrary.wiley.com/doi/10.1111/j.1600-0668.2007.00475.x/pdf]

IOM, 2011: *Climate Change, the Indoor Environment, and Health*. The National Academies Press. [Available online at www.nap.edu]

Mudarri, D., and W. J. Fisk, 2007: Public health and economic impact of dampness and mold. *Indoor Air*, **17**, 226-235, doi:10.1111/j.1600-0668.2007.00474.x. [Available online at http://onlinelibrary.wiley.com/doi/10.1111/j.1600-0668.2007.00474.x/pdf]

Wolf, J., N. R. R. O'Neill, C.A., M. T. Muilenberg, and L. H. Ziska, 2010: Elevated atmospheric carbon dioxide concentrations amplify *Alternaria alternata* sporulation and total antigen production. *Environmental Health Perspectives*, **118**, 1223-1228, doi:10.1289/ehp.0901867.

21. Curriero, F. C., J. A. Patz, J. B. Rose, and S. Lele, 2001: The association between extreme precipitation and waterborne disease outbreaks in the United States, 1948–1994. *American Journal of Public Health*, **91**, 1194-1199, doi:10.2105/AJPH.91.8.1194.

ECDC, 2012: Assessing the potential impacts of climate change on food- and waterborne diseases in Europe. European Centre for Disease Prevention and Control, Stockholm. [Available online at http://www.ecdc.europa.eu/en/publications/Publications/1203-TER-Potential-impacts-climate-change-food-water-borne-diseases.pdf]

Semenza, J. C., J. E. Suk, V. Estevez, K. L. Ebi, and E. Lindgren, 2011: Mapping climate change vulnerabilities to infectious diseases in Europe. *Environmental Health Perspectives*, **120**, 385-392, doi:10.1289/ehp.1103805. [Available online at http://www.ncbi.nlm.nih.gov/pmc/articles/PMC3295348/pdf/ehp.1103805.pdf]

22. Fleury, M., D. F. Charron, J. D. Holt, O. B. Allen, and A. R. Maarouf, 2006: A time series analysis of the relationship of ambient temperature and common bacterial enteric infections in two Canadian provinces. *International Journal of Biometeorology*, **50**, 385-391, doi:10.1007/s00484-006-0028-9.

Hall, G. V., I. C. Hanigan, K. D. G. Dear, and H. Vally, 2011: The influence of weather on community gastroenteritis in Australia. *Epidemiology and Infection*, **139**, 927-936, doi:10.1017/s0950268810001901.

Hu, W., K. Mengersen, S.-Y. Fu, and S. Tong, 2010: The use of ZIP and CART to model cryptosporidiosis in relation to climatic variables. *International Journal of Biometeorology*, **54**, 433-440, doi:10.1007/s00484-009-0294-4.

Hu, W., S. Tong, K. Mengersen, and D. Connell, 2007: Weather variability and the incidence of cryptosporidiosis: Comparison of time series Poisson regression and SARIMA models. *Annals of Epidemiology*, **17**, 679-688, doi:10.1016/j.annepidem.2007.03.020.

Lipp, E. K., A. Huq, R. R. Colwell, E. K. Lipp, A. Huq, and R. R. Colwell, 2002: Effects of global climate on infectious disease: The cholera model. *Clinical Microbiology Reviews*, **15**, 757-770, doi:10.1128/CMR.15.4.757-770.2002. [Available online at http://cmr.asm.org/content/15/4/757.full.pdf+html]

Naumova, E. N., J. S. Jagai, B. Matyas, A. DeMaria, I. B. MacNeill, and J. K. Griffiths, 2007: Seasonality in six enterically transmitted diseases and ambient temperature. *Epidemiology and Infection*, **135**, 281-292, doi:10.1017/S0950268806006698. [Available online at http://www.ncbi.nlm.nih.gov/pmc/articles/PMC2870561/]

Onozuka, D., M. Hashizume, and A. Hagihara, 2010: Effects of weather variability on infectious gastroenteritis. *Epidemiology and Infection*, **138**, 236-243, doi:10.1017/s0950268809990574.

23. Febriani, Y., P. Levallois, S. Gingras, P. Gosselin, S. E. Majowicz, and M. D. Fleury, 2010: The association between farming activities, precipitation, and the risk of acute gastrointestinal illness in rural municipalities of Quebec, Canada: A cross-sectional study. *BMC Public Health*, **10**, 48, doi:10.1186/1471-2458-10-48. [Available online at http://www.biomedcentral.com/content/pdf/1471-2458-10-48.pdf]

Nichols, G., C. Lane, N. Asgari, N. Q. Verlander, and A. Charlett, 2009: Rainfall and outbreaks of drinking water related disease and in England and Wales. *Journal of Water Health*, **7**, 1-8, doi:10.2166/

wh.2009.143. [Available online at http://www.iwaponline.com/jwh/007/0001/0070001.pdf]

24. Harper, S. L., V. L. Edge, C. J. Schuster-Wallace, O. Berke, and S. A. McEwen, 2011: Weather, water quality and infectious gastrointestinal illness in two Inuit communities in Nunatsiavut, Canada: Potential implications for climate change. *Ecohealth*, **8**, 93-108, doi:10.1007/s10393-011-0690-1.

25. Rizak, S., and S. E. Hrudey, 2008: Drinking-water safety: Challenges for community-managed systems. *Journal of Water Health*, **6**, 33-42, doi:10.2166/wh.2008.033. [Available online at http://www.iwaponline.com/jwh/006/s033/006s033.pdf]

26. Baker-Austin, C., J. A. Trinanes, N. G. H. Taylor, R. Hartnell, A. Siitonen, and J. Martinez-Urtaza, 2012: Emerging *Vibrio* risk at high latitudes in response to ocean warming. *Nature Climate Change*, **3**, 73-77, doi:10.1038/nclimate1628.

CDC, 1998: Outbreak of *Vibrio parahaemolyticus* infections associated with eating raw oysters-Pacific Northwest, 1997. Centers for Disease Control and Prevention. *Morbidity and Mortality Weekly Report*, **47**, 457-462. [Available online at http://www.cdc.gov/mmwr/preview/mmwrhtml/00053377.htm]

Patz, J. A., S. J. Vavrus, C. K. Uejio, and S. L. McLellan, 2008: Climate change and waterborne disease risk in the Great Lakes region of the US. *American Journal of Preventive Medicine*, **35**, 451-458, doi:10.1016/j.amepre.2008.08.026. [Available online at http://www.ajpmonline.org/article/S0749-3797(08)00702-2/fulltext]

27. Patz, J., D. Campbell-Lendrum, H. Gibbs, and R. Woodruff, 2008: Health impact assessment of global climate change: Expanding on comparative risk assessment approaches for policy making. *Annual Review of Public Health*, **29**, 27-39, doi:10.1146/annurev.publhealth.29.020907.090750. [Available online at http://www.sage.wisc.edu/pubs/articles/M-Z/patz/PatzAnnRevPubHealth2008.pdf]

28. NWS, cited 2012: Weather Fatalities. [Available online at http://www.nws.noaa.gov/os/hazstats/resources/weather_fatalities.pdf]

29. Åström, D. O., F. Bertil, and R. Joacim, 2011: Heat wave impact on morbidity and mortality in the elderly population: A review of recent studies. *Maturitas*, **69**, 99-105, doi:10.1016/j.maturitas.2011.03.008.

Huang, C., A. G. Barnett, X. Wang, P. Vaneckova, G. FitzGerald, and S. Tong, 2011: Projecting future heat-related mortality under climate change scenarios: A systematic review. *Environmental Health Perspectives*, **119**, 1681-1690, doi:10.1289/Ehp.1103456. [Available online at http://ehp.niehs.nih.gov/wp-content/uploads/119/12/ehp.1103456.pdf]

Li, B., S. Sain, L. O. Mearns, H. A. Anderson, S. Kovats, K. L. Ebi, M. Y. V. Bekkedal, M. S. Kanarek, and J. A. Patz, 2012: The impact of extreme heat on morbidity in Milwaukee, Wisconsin. *Climatic Change*, **110**, 959-976, doi:10.1007/s10584-011-0120-y.

Ye, X., R. Wolff, W. Yu, P. Vaneckova, X. Pan, and S. Tong, 2012: Ambient temperature and morbidity: A review of epidemiological evidence. *Environmental Health Perspectives*, **120**, 19-28, doi:10.1289/ehp.1003198.

Zanobetti, A., M. S. O'Neill, C. J. Gronlund, and J. D. Schwartz, 2012: Summer temperature variability and long-term survival among elderly people with chronic disease. *Proceedings of the National Academy of Sciences*, **109**, 6608-6613, doi:10.1073/pnas.1113070109.

30. Basu, R., 2009: High ambient temperature and mortality: A review of epidemiologic studies from 2001 to 2008. *Environmental Health*, **8**, 1-13, doi:10.1186/1476-069X-8-40.

31. Rey, G., E. Jougla, A. Fouillet, G. Pavillon, P. Bessemoulin, P. Frayssinet, J. Clavel, and D. Hémon, 2007: The impact of major heat waves on all-cause and cause-specific mortality in France from 1971 to 2003. *International Archives of Occupational and Environmental Health*, **80**, 615-626, doi:10.1007/s00420-007-0173-4.

32. Knowlton, K., M. Rotkin-Ellman, G. King, H. G. Margolis, D. Smith, G. Solomon, R. Trent, and P. English, 2009: The 2006 California heat wave: Impacts on hospitalizations and emergency department visits. *Environmental Health Perspectives*, **117**, 61-67, doi:10.1289/ehp.11594. [Available online at http://www.ncbi.nlm.nih.gov/pmc/articles/PMC2627866/pdf/EHP-117-61.pdf]

Lin, S., M. Luo, R. J. Walker, X. Liu, S. A. Hwang, and R. Chinery, 2009: Extreme high temperatures and hospital admissions for respiratory and cardiovascular diseases. *Epidemiology*, **20**, 738-746, doi:10.1097/EDE.0b013e3181ad5522.

Nitschke, M., G. R. Tucker, A. L. Hansen, S. Williams, Y. Zhang, and P. Bi, 2011: Impact of two recent extreme heat episodes on morbidity and mortality in Adelaide, South Australia: A case-series analysis. *Environmental Health*, **10**, 1-9, doi:10.1186/1476-069X-10-42. [Available online at http://www.biomedcentral.com/content/pdf/1476-069X-10-42.pdf]

Ostro, B. D., L. A. Roth, R. S. Green, and R. Basu, 2009: Estimating the mortality effect of the July 2006 California heat wave. *Environmental Research*, **109**, 614-619, doi:10.1016/j.envres.2009.03.010. [Available online at http://www.energy.ca.gov/2009publications/CEC-500-2009-036/CEC-500-2009-036-F.PDF]

33. Gershunov, A., D. R. Cayan, and S. F. Iacobellis, 2009: The great 2006 heat wave over California and Nevada: Signal of an increasing trend. *Journal of Climate*, **22**, 6181-6203, doi:10.1175/2009jcli2465.1. [Available online at http://journals.ametsoc.org/doi/pdf/10.1175/2009JCLI2465.1]

Gershunov, A., Z. Johnston, H. G. Margolis, and K. Guirguis, 2011: The California heat wave 2006 with impacts on statewide medical emergency: A space-time analysis. *Geography Research Forum*, **31**, 6-31.

Sheridan, S., C. Lee, M. Allen, and L. Kalkstein, 2011: A Spatial Synoptic Classification Approach to Projected Heat Vulnerability in California Under Future Climate Change Scenarios. Final Report to the California Air Resources Board. Research Contract 07-304, 155 pp., California Air Resources Board and the California Environmental Protection Agency. [Available online at http://www.arb.ca.gov/newsrel/2011/HeatImpa.pdf]

Sheridan, S. C., M. J. Allen, C. C. Lee, and L. S. Kalkstein, 2012: Future heat vulnerability in California, Part II: Projecting future heat-related mortality. *Climatic Change*, **115**, 311-326, doi:10.1007/s10584-012-0437-1.

Sheridan, S. C., C. C. Lee, M. J. Allen, and L. S. Kalkstein, 2012: Future heat vulnerability in California, Part I: Projecting future weather types and heat events. *Climatic Change*, **115**, 291-309, doi:10.1007/s10584-012-0436-2.

34. Barnett, A. G., 2007: Temperature and cardiovascular deaths in the US elderly: Changes over time. *Epidemiology*, **18**, 369-372, doi:10.1097/01.ede.00002575.15.34445.a0.

Kalkstein, L. S., S. Greene, D. M. Mills, and J. Samenow, 2011: An evaluation of the progress in reducing heat-related human mortality in major US cities. *Natural Hazards*, **56**, 113-129, doi:10.1007/s11069-010-9552-3.

35. Johnson, D. P., J. S. Wilson, and G. C. Luber, 2009: Socioeconomic indicators of heat-related health risk supplemented with remotely sensed data. *International Journal of Health Geographics*, **8**, 1-13, doi:10.1186/1476-072X-8-57. [Available online at http://www.ij-healthgeographics.com/content/8/1/57]

Wilby, R. L., 2008: Constructing climate change scenarios of urban heat island intensity and air quality. *Environment and Planning B: Planning and Design*, **35**, 902-919, doi:10.1068/b33066t.

CDC, 2012: Heat-related deaths after an extreme heat event — four states, 2012, and United States, 1999–2009. Centers for Disease Control and Prevention. *Morbidity and Mortality Weekly Report*, **62**, 433-436. [Available online at http://www.cdc.gov/mmwr/preview/mmwrhtml/mm6222a1.htm?s_cid=mm6222a1_w]

36. Medina-Ramón, M., and J. Schwartz, 2007: Temperature, temperature extremes, and mortality: A study of acclimatisation and effect modification in 50 US cities. *Occupational and Environmental Medicine*, **64**, 827-833, doi:10.1136/oem.2007.033175. [Available online at http://oem.bmj.com/content/64/12/827.full.pdf+html]

Yu, W., K. Mengersen, X. Wang, X. Ye, Y. Guo, X. Pan, and S. Tong, 2011: Daily average temperature and mortality among the elderly: A meta-analysis and systematic review of epidemiological evidence. *International Journal of Biometeorology*, **56**, 569-581, doi:10.1007/s00484-011-0497-3.

Li, T., R. M. Horton, and P. L. Kinney, 2013: Projections of seasonal patterns in temperature-related deaths for Manhattan, New York. *Nature Climate Change*, **3**, 717-721, doi:10.1038/nclimate1902.

37. Epstein, P., 2010: The ecology of climate change and infectious diseases: Comment. *Ecology*, **91**, 925-928, doi:10.1890/09-0761.1.

Reiter, P., 2008: Climate change and mosquito-borne disease: Knowing the horse before hitching the cart. *Revue Scientifique et Technique-Office International des Épizooties*, **27**, 383-398. [Available online at http://ocean.otr.usm.edu/~w777157/Reiter%202008.pdf]

Rosenthal, J., 2009: Climate change and the geographic distribution of infectious diseases. *EcoHealth*, **6**, 489-495, doi:10.1007/s10393-010-0314-1. [Available online at http://download.springer.com/static/pdf/305/art%253A10.1007%252Fs10393-010-0314-1.pdf?auth66=1362580261_e7030052d90896d4fec0fbabe27e8083&ext=.pdf]

Russell, R. C., 2009: Mosquito-borne disease and climate change in Australia: Time for a reality check. *Australian Journal of Entomology*, **48**, 1-7, doi:10.1111/j.1440-6055.2000.00677.x.

Tabachnick, W. J., 2010: Challenges in predicting climate and environmental effects on vector-borne disease episystems in a changing world. *Journal of Experimental Biology*, **213**, 946-954, doi:10.1242/jeb.037564. [Available online at http://jeb.biologists.org/content/213/6/946.long]

Rogers, D. J., and S. E. Randolph, 2006: Climate change and vector-borne diseases. *Advances in Parasitology*, S. I. Hay, A. Graham, and D. J. Rogers, Eds., Academic Press, 345-381.

Ogden, N. H., L. R. Lindsay, M. Morshed, P. N. Sockett, and H. Artsob, 2009: The emergence of Lyme disease in Canada. *Canadian Medical Association Journal*, **180**, 1221-1224, doi:10.1503/cmaj.080148. [Available online at http://www.cmaj.ca/content/180/12/1221.full]

38. Gage, K. L., T. R. Burkot, R. J. Eisen, and E. B. Hayes, 2008: Climate and vectorborne diseases. *American Journal of Preventive Medicine*, **35**, 436-450, doi:10.1016/j.amepre.2008.08.030.

Lafferty, K. D., 2009: The ecology of climate change and infectious diseases. *Ecology*, **90**, 888-900, doi:10.1890/08-0079.1.

39. McGregor, G. R., 2011: Human biometeorology. *Progress in Physical Geography*, **36**, 93-109, doi:10.1177/0309133311417942.

40. Mills, J. N., K. L. Gage, and A. S. Khan, 2010: Potential influence of climate change on vector-borne and zoonotic diseases: A review and proposed research plan. *Environmental Health Perspectives*, **118**, 1507-1514, doi:10.1289/ehp.0901389. [Available online at http://www.ncbi.nlm.nih.gov/pmc/articles/PMC2974686/]

41. Hess, J. J., J. Z. McDowell, and G. Luber, 2012: Integrating climate change adaptation into public health practice: Using adaptive management to increase adaptive capacity and build resilience. *Environmental Health Perspectives*, **120**, 171-179, doi:10.1289/ehp.1103515. [Available online at http://www.ncbi.nlm.nih.gov/pmc/articles/PMC3279431/]

Patz, J. A., and M. B. Hahn, 2013: Climate change and human health: A one health approach. *Current Topics in Microbiology and Immunology*, Springer Berlin Heidelberg, 141-171.

42. Wilson, K., 2009: Climate change and the spread of infectious ideas. *Ecology*, **90**, 901-902, doi:10.1890/08-2027.1.

43. Diuk-Wasser, M. A., G. Vourc'h, P. Cislo, A. G. Hoen, F. Melton, S. A. Hamer, M. Rowland, R. Cortinas, G. J. Hickling, J. I. Tsao, A. G. Barbour, U. Kitron, J. Piesman, and D. Fish, 2010: Field and climate-based model for predicting the density of host-king nymphal *Ixodes scapularis*, an important vector of tick-borne disease agents in the eastern United States. *Global Ecology and Biogeography*, **19**, 504-514, doi:10.1111/j.1466-8238.2010.00526.x.

Keesing, F., J. Brunner, S. Duerr, M. Killilea, K. LoGiudice, K. Schmidt, H. Vuong, and R. S. Ostfeld, 2009: Hosts as ecological traps for the vector of Lyme disease. *Proceedings of the Royal Society B: Biological Sciences*, **276**, 3911-3919, doi:10.1098/rspb.2009.1159.

Ogden, N. H., L. St-Onge, I. K. Barker, S. Brazeau, M. Bigras-Poulin, D. F. Charron, C. Francis, A. Heagy, L. R. Lindsay, A. Maarouf, P. Michel, F. Milord, C. J. O'Callaghan, L. Trudel, and R. A. Thompson, 2008: Risk maps for range expansion of the Lyme disease vector, *Ixodes scapularis*, in Canada now and with climate change. *International Journal of Health Geographics*, **1**, 24, doi:10.1186/1476-072X-7-24. [Available online at http://www.ij-healthgeographics.com/content/7/1/24]

CDC, cited 2013: Interactive Lyme Disease Map. Centers for Disease Control and Prevention. [Available online at http://www.cdc.gov/lyme/stats/maps/interactiveMaps.html]

44. Degallier, N., C. Favier, C. Menkes, M. Lengaigne, W. M. Ramalho, R. Souza, J. Servain, and J. P. Boulanger, 2010: Toward an early warning system for dengue prevention: Modeling climate impact on dengue transmission. *Climatic Change*, **98**, 581-592, doi:10.1007/s10584-009-9747-3. [Available online at http://www.locean-ipsl.upmc.fr/~ndelod/production/climatic_Change.pdf]

Johansson, M. A., D. A. T. Cummings, and G. E. Glass, 2009: Multiyear climate variability and dengue—El Niño southern oscillation, weather, and dengue incidence in Puerto Rico, Mexico, and Thailand: A longitudinal data analysis. *PLoS Medicine*, **6**, e1000168, doi:10.1371/journal.pmed.1000168. [Available online at http://www.plosmedicine.org/article/info%3Adoi%2F10.1371%2Fjournal.pmed.1000168]

Jury, M. R., 2008: Climate influence on dengue epidemics in Puerto Rico. *International Journal of Environmental Health Research*, **18**, 323-334, doi:10.1080/09603120701849836.

Kolivras, K. N., 2010: Changes in dengue risk potential in Hawaii, USA, due to climate variability and change. *Climate Research*, **42**, 1-11, doi:10.3354/cr00861. [Available online at http://www.int-res.com/articles/cr2010/42/c042p001.pdf]

Lambrechts, L., K. P. Paaijmans, T. Fansiri, L. B. Carrington, L. D. Kramer, M. B. Thomas, and T. W. Scott, 2011: Impact of daily temperature fluctuations on dengue virus transmission by *Aedes aegypti*. *Proceedings of the National Academy of Sciences*, **108**, 7460-7465, doi:10.1073/pnas.1101377108. [Available online at http://www.pnas.org/content/108/18/7460.full.pdf+html]

Ramos, M. M., H. Mohammed, E. Zielinski-Gutierrez, M. H. Hayden, J. L. R. Lopez, M. Fournier, A. R. Trujillo, R. Burton, J. M. Brunkard, L. Anaya-Lopez, A. A. Banicki, P. K. Morales, B. Smith, J. L. Muñoz, and S. H. Waterman, 2008: Epidemic dengue and dengue hemorrhagic fever at the Texas–Mexico border: Results of a household-based seroepidemiologic survey, December 2005. *The American Journal of Tropical Medicine and Hygiene*, **78**, 364-369. [Available online at http://www.ajtmh.org/content/78/3/364.full.pdf+html]

Gong, H., A. T. DeGaetano, and L. C. Harrington, 2011: Climate-based models for West Nile *Culex* mosquito vectors in the Northeastern U.S. *International Journal of Biometeorology*, **55**, 435-446, doi:10.1007/s00484-010-0354-9.

Morin, C. W., and A. C. Comrie, 2010: Modeled response of the West Nile virus vector *Culex quinquefasciatus* to changing climate using the dynamic mosquito simulation model. *International Journal of Biometeorology*, **54**, 517-529, doi:10.1007/s00484-010-0349-6.

CDC, cited 2012: Rocky Mountain Spotted Fever. Centers for Disease Control and Prevention. [Available online at www.cdc.gov/rmsf/stats]

Nakazawa, Y., R. Williams, A. T. Peterson, P. Mead, E. Staples, and K. L. Gage, 2007: Climate change effects on plague and tularemia in the United States. *Vector-Borne and Zoonotic Diseases*, **7**, 529-540, doi:10.1089/vbz.2007.0125.

45. Ogden, N. H., C. Bouchard, K. Kurtenbach, G. Margos, L. R. Lindsay, L. Trudel, S. Nguon, and F. Milord, 2010: Active and passive surveillance and phylogenetic analysis of *Borrelia burgdorferi* elucidate the process of Lyme disease risk emergence in Canada. *Environmental Health Perspectives*, **118**, 909-914, doi:10.1289/

ehp.0901766. [Available online at http://www.ncbi.nlm.nih.gov/pmc/articles/PMC2920908/pdf/ehp-118-909.pdf]

46. Brownstein, J. S., T. R. Holford, and D. Fish, 2005: Effect of climate change on Lyme disease risk in North America. *EcoHealth*, **2**, 38-46, doi:10.1007/s10393-004-0139-x.

47. Grabow, M. L., S. N. Spak, T. Holloway, B. Stone Jr, A. C. Mednick, and J. A. Patz, 2012: Air quality and exercise-related health benefits from reduced car travel in the midwestern United States. *Environmental Health Perspectives*, **120**, 68-76, doi:10.1289/ehp.1103440. [Available online at http://www.ncbi.nlm.nih.gov/pmc/articles/PMC3261937/pdf/ehp.1103440.pdf]

Haines, A., A. J. McMichael, K. R. Smith, I. Roberts, J. Woodcock, A. Markandya, B. G. Armstrong, D. Campbell-Lendrum, A. D. Dangour, M. Davies, N. Bruce, C. Tonne, M. Barrett, and P. Wilkinson, 2009: Public health benefits of strategies to reduce greenhouse-gas emissions: Overview and implications for policy makers. *The Lancet*, **374**, 2104-2114, doi:10.1016/s0140-6736(09)61759-1. [Available online at http://www.sciencedirect.com/science/article/pii/S0140673609617591]

Maizlish, N., J. Woodcock, S. Co, B. Ostro, A. Fanai, and D. Fairley, 2013: Health cobenefits and transportation-related reductions in greenhouse gas emissions in the San Francisco Bay area. *American Journal of Public Health*, **103**, 703-709, doi:10.2105/ajph.2012.300939. [Available online at http://www.cdph.ca.gov/programs/CCDPHP/Documents/ITHIM_Technical_Report11-21-11.pdf]

48. Bambrick, H. J., A. G. Capon, G. B. Barnett, R. M. Beaty, and A. J. Burton, 2011: Climate change and health in the urban environment: Adaptation opportunities in Australian cities. *Asia-Pacific Journal of Public Health*, **23**, 67S-79S, doi:10.1177/1010539510391774.

49. Wilkinson, P., K. R. Smith, S. Beevers, C. Tonne, and T. Oreszczyn, 2007: Energy, energy efficiency, and the built environment. *The Lancet*, **370**, 1175-1187, doi:10.1016/S0140-6736(07)61255-0.

50. Stone, B., J. J. Hess, and H. Frumkin, 2010: Urban form and extreme heat events: Are sprawling cities more vulnerable to climate change than compact cities? *Environmental Health Perspectives*, **118**, 1425-1428, doi:10.1289/ehp.0901879. [Available online at http://www.ncbi.nlm.nih.gov/pmc/articles/PMC2957923/pdf/ehp-118-1425.pdf]

Finding 6: Infrastructure

1. NRC, 2010: Adapting to Impacts of Climate Change. America's Climate Choices: Report of the Panel on Adapting to the Impacts of Climate Change. National Research Council. The National Academies Press, 292 pp. [Available online at http://www.nap.edu/catalog.php?record_id=12783]

2. Rosenzweig, C., W. Solecki, S. A. Hammer, and S. Mehrotra, 2010: Cities lead the way in climate-change action. *Nature*, **467**, 909-911, doi:10.1038/467909a. [Available online at http://ccrun.org/sites/ccrun/files/attached_files/2010_Rosenzweig_etal_2.pdf]

3. Foster, J., S. Winkelman, and A. Lowe, 2011: Lessons Learned on Local Climate Adaptation from the Urban Leaders Adaptation Initiative, 23 pp., The Center for Clean Air Policy, Washington, D.C. [Available online at http://www.ccap.org/docs/resources/988/Urban_Leaders_Lessons_Learned_FINAL.pdf]

4. Jacob, K., C. Rosenzweig, R. Horton, D. Major, and V. Gornitz, 2008: MTA Adaptations to Climate Change: A Categorical

Imperative State agency report, 48 pp., State of New York, Metropolitan Transportation Authority, New York, NY. [Available online at http://www.mta.info/sustainability/pdf/Jacob_et%20 al_MTA_Adaptation_Final_0309.pdf]

New York State, 2011: Responding to Climate Change in New York State: The ClimAID Integrated Assessment for Effective Climate Change Adaptation in New York State. Vol. 1244, Blackwell Publishing, 649 pp. [Available online at http://onlinelibrary.wiley. com/doi/10.1111/j.1749-6632.2011.06331.x/pdf]

New York State Sea Level Rise Task Force, 2010: Report to the Legislature, 102 pp., New York State Department of Environmental Conservation, Albany, NY. [Available online at http://www.dec. ny.gov/docs/administration_pdf/slrtffinalrep.pdf]

Zimmerman, R., and C. Faris, 2010: Infrastructure impacts and adaptation challenges. Annals of the New York Academy of Sciences, **1196**, 63-86, doi:10.1111/j.1749-6632.2009.05318.x.

5. EPA, 2011: Inventory of U.S. Greenhouse Gas Emissions and Sinks: 2000 – 2009. EPA 430-R-11-005, 459 pp., U.S. Environmental Protection Agency, Washington, D.C. [Available online at http:// www.epa.gov/climatechange/Downloads/ghgemissions/US-GHG-Inventory-2011-complete_Report.pdf]

6. Kafalenos, R. S., K. J. Leonard, D. M. Beagan, V. R. Burkett, B. D. Keim, A. Meyers, D. T. Hunt, R. C. Hyman, M. K. Maynard, B. Fritsche, R. H. Henk, E. J. Seymour, L. E. Olson, J. R. Potter, and M. J. Savonis, 2008: Ch. 4: What are the implications of climate change and variability for Gulf Coast transportation? *Impacts of Climate Change and Variability on Transportation Systems and Infrastructure: Gulf Study, Phase I. A report by the U.S. Climate Change Science Program and the Subcommittee on Global Change Research. Final Report of Synthesis and Assessment Product 4.7*, M. J. Savonis, V. R. Burkett, and J. R. Potter, Eds., U.S. Department of Transportation. [Available online at http://www.climatescience.gov/Library/sap/sap4-7/final-report/ sap4-7-final-all.pdf]

7. Wilbanks, T., D. Bilello, D. Schmalzer, and M. Scott, 2012: Climate Change and Energy Supply and Use. Technical Report to the U.S. Department of Energy in Support of the National Climate Assessment, 79 pp., Oak Ridge National Laboratory, U.S. Department of Energy, Office of Science, Oak Ridge, TN. [Available online at http://www.esd.ornl.gov/eess/EnergySupplyUse.pdf]

8. NCDC, cited 2012: Heating & Cooling Degree Day Data. NOAA's National Climatic Data Center. [Available online at http://www. ncdc.noaa.gov/oa/documentlibrary/hcs/hcs.html]

Finding 7: Water

1. Pietrowsky, R., D. Raff, C. McNutt, M. Brewer, T. Johnson, T. Brown, M. Ampleman, C. Baranowski, J. Barsugli, L. D. Brekke, L. Brekki, M. Crowell, D. Easterling, A. Georgakakos, N. Gollehon, J. Goodrich, K. A. Grantz, E. Greene, P. Groisman, R. Heim, C. Luce, S. McKinney, R. Najjar, M. Nearing, D. Nover, R. Olsen, C. Peters-Lidard, L. Poff, K. Rice, B. Rippey, M. Rodgers, A. Rypinski, M. Sale, M. Squires, R. Stahl, E. Z. Stakhiv, and M. Strobel, 2012: Water Resources Sector Technical Input Report in Support of the U.S. Global Change Research Program, National Climate Assessment - 2013, 31 pp.

2. Kenny, J. F., N. L. Barber, S. S. Hutson, K. S. Linsey, J. K. Lovelace, and M. A. Maupin, 2009: Estimated Use of Water in the United States in 2005. U.S. Geological Survey Circular 1344, 52 pp., U.S. Geological Survey Reston, VA. [Available online at http://pubs. usgs.gov/circ/1344/]

3. Averyt, K., J. Fisher, A. Huber-Lee, A. Lewis, J. Macknick, N. Madden, J. Rogers, and S. Tellinghuisen, 2011: Freshwater Use by US Power Plants: Electricity's Thirst for a Precious Resource. A Report of the Energy and Water in a Warming World initiative, 62 pp., Union of Concerned Scientists. [Available online at http://www. ucsusa.org/assets/documents/clean_energy/ew3/ew3-freshwater-use-by-us-power-plants.pdf]

4. Brown, T. C., R. Foti, and J. A. Ramirez, 2013: Projecting fresh water withdrawals in the United States under a changing climate. *Water Resources Research*, **49**, 1259-1276, doi:10.1002/wrcr.20076. [Available online at http://onlinelibrary.wiley.com/doi/10.1002/wrcr.20076/ pdf]

5. Pruski, F. F., and M. A. Nearing, 2002: Climate-induced changes in erosion during the 21st century for eight U.S. locations. *Water Resources Research*, **38**, 34-31 - 34-11, doi:10.1029/2001WR000493. [Available online at http://onlinelibrary.wiley.com/ doi/10.1029/2001WR000493/pdf]

——, 2002: Runoff and soil-loss responses to changes in precipitation: A computer simulation study. *Journal of Soil and Water Conservation*, **57**, 7-16.

6. Osterkamp, W. R., and C. R. Hupp, 2010: Fluvial processes and vegetation--Glimpses of the past, the present, and perhaps the future. *Geomorphology*, **116**, 274-285, doi:10.1016/j.geomorph.2009.11.018.

7. Roy, S. B., L. Chen, E. H. Girvetz, E. P. Maurer, W. B. Mills, and T. M. Grieb, 2012: Projecting water withdrawal and supply for future decades in the U.S. under climate change scenarios. *Environmental Science & Technology*, **46**, 2545–2556, doi:10.1021/es2030774.

8. Skaggs, R., T. C. Janetos, K. A. Hibbard, and J. S. Rice, 2012: Climate and Energy-Water-Land System Interactions Technical Report to the U.S. Department of Energy in Support of the National Climate Assessment, 152 pp., Pacific Northwest National Laboratory, Richland, Washington. [Available online at http://climatemodeling. science.energy.gov/f/PNNL-21185_FINAL_REPORT.pdf]

Finding 8: Agriculture

1. Mishra, V., and K. A. Cherkauer, 2010: Retrospective droughts in the crop growing season: Implications to corn and soybean yield in the midwestern United States. *Agricultural and Forest Meteorology*, **150**, 1030-1045, doi:10.1016/j.agrformet.2010.04.002.

2. Changnon, S. A., and D. Winstanley, 1999: *Long-Term Variations in Seasonal Weather Conditions and Their Impacts on Crop Production and Water Resources in Illinois. Water Survey Research Report RR-127*. Illinois State Water Survey, Dept. of Natural Resources. [Available online at http://www.isws.uiuc.edu/pubdoc/RR/ISWSRR-127.pdf]

3. Karl, T. R., B. E. Gleason, M. J. Menne, J. R. McMahon, R. R. Helm, Jr., M. J. Brewer, K. E. Kunkel, D. S. Arndt, J. L. Privette, J. J. Bates, P. Y. Groisman, and D. R. Easterling, 2012: U.S. temperature and drought: Recent anomalies and trends. *Eos, Transactions, American Geophysical Union*, **93**, 473-474, doi:10.1029/2012EO470001. [Available online at http://onlinelibrary.wiley.com/doi/10.1029/2012EO470001/pdf]

4. Kunkel, K. E., 1989: A surface energy budget view of the 1988 midwestern United States drought. *Boundary-Layer Meteorology*, **48**, 217-225, doi:10.1007/BF00158325.

5. U.S. Census Bureau, 2012: The 2012 Statistical Abstract: Agriculture, 533-558 pp., U.S. Census Bureau, U.S. Department of Commerce, Washington, D.C. [Available online at http://www.census.gov/prod/2011pubs/12statab/agricult.pdf]

6. Colby, B., and P. Tanimoto, 2011: Using climate information to improve electric utility load forecasting. *Adaptation and Resilience: The Economics of Climate-Water-Energy Challenges in the Arid Southwest*, B. G. Colby, and G. B. Frisvold, Eds., RFF Press, 207-228.

7. Mader, T. L., 2012: Impact of environmental stress on feedlot cattle. *Western Section, American Society of Animal Science*, **62**, 335-339.

Finding 9: Indigenous Peoples

1. Maynard, N. G., Ed., 2002: Native Peoples-Native Homelands Climate Change Workshop. Final Report: Circles of Wisdom. Albuquerque Convention Center, Albuquerque, New Mexico, NASA Goddard Space Flight Center. [Available online at http://www.usgcrp.gov/usgcrp/Library/nationalassessment/native.pdf]

   Houser, S., V. Teller, M. MacCracken, R. Gough, and P. Spears, 2001: Ch. 12: Potential consequences of climate variability and change for native peoples and homelands. *Climate Change Impacts in the United States: Potential Consequences of Climate Change and Variability and Change*, Cambridge University Press, 351-377. [Available online at http://www.gcrio.org/NationalAssessment/12NA.pdf]

2. Cochran, P., O. H. Huntington, C. Pungowiyi, S. Tom, F. S. Chapin, III, H. P. Huntington, N. G. Maynard, and S. F. Trainor, 2013: Indigenous frameworks for observing and responding to climate change in Alaska. *Climatic Change*, **120**, 557-567, doi:10.1007/s10584-013-0735-2.

3. Lynn, K., J. Daigle, J. Hoffman, F. Lake, N. Michelle, D. Ranco, C. Viles, G. Voggesser, and P. Williams, 2013: The impacts of climate change on tribal traditional foods. *Climatic Change*, **120**, 545-556, doi:10.1007/s10584-013-0736-1.

   Maldonado, J. K., C. Shearer, R. Bronen, K. Peterson, and H. Lazrus, 2013: The impact of climate change on tribal communities in the US: Displacement, relocation, and human rights. *Climatic Change*, **120**, 601-614, doi:10.1007/s10584-013-0746-z.

   Voggesser, G., K. Lynn, J. Daigle, F. K. Lake, and D. Ranco, 2013: Cultural impacts to tribes from climate change influences on forests. *Climatic Change*, **120**, 615-626, doi:10.1007/s10584-013-0733-4.

4. Redsteer, M. H., K. B. Kelley, H. Francis, and D. Block, 2011: Disaster Risk Assessment Case Study: Recent Drought on the Navajo Nation, southwestern United States. Contributing Paper for the Global Assessment Report on Disaster Risk Reduction, 19 pp., United Nations Office for Disaster Risk Reduction and U.S. Geological Survey, Reston, VA. [Available online at http://www.preventionweb.net/english/hyogo/gar/2011/en/bgdocs/Redsteer_Kelley_Francis_&_Block_2010.pdf]

5. Christensen, K., 2003: Cooperative Drought Contingency Plan - Hualapai Reservation. Hualapai Tribe Department of Natural Resources, Peach Springs, AZ. [Available online at http://hualapai.org/resources/Aministration/droughtplan.rev3BOR.pdf]

Ferguson, D. B., C. Alvord, M. Crimmins, M. Hiza Redsteer, M. Hayes, C. McNutt, R. Pulwarty, and M. Svoboda, 2011: Drought Preparedness for Tribes in the Four Corners Region. Report from April 2010 Workshop. Tucson, AZ: Climate Assessment for the Southwest., 42 pp., The Climate Assessment for the Southwest (CLIMAS), The Institute of the Environment, The University of Arizona [Available online at http://www.drought.gov/workshops/tribal/Drought-Preparedness-Tribal-Lands-FoursCorners-2011-1.pdf]

Garfin, G., A. Jardine, R. Merideth, M. Black, and S. LeRoy, Eds., 2013: *Assessment of Climate Change in the Southwest United States: A Report Prepared for the National Climate Assessment*. Island press, 528 pp. [Available online at http://swccar.org/sites/all/themes/files/SW-NCA-color-FINALweb.pdf]

Gautam, M. R., K. Chief, and W. J. Smith, Jr., 2013: Climate change in arid lands and Native American socioeconomic vulnerability: The case of the Pyramid Lake Paiute Tribe. *Climatic Change*, **120**, 585-599, doi:10.1007/s10584-013-0737-0. [Available online at http://link.springer.com/content/pdf/10.1007%2Fs10584-013-0737-0.pdf]

6. Dittmer, K., 2013: Changing streamflow on Columbia basin tribal lands—climate change and salmon. *Climatic Change*, **120**, 627-641, doi:10.1007/s10584-013-0745-0. [Available online at http://link.springer.com/content/pdf/10.1007%2Fs10584-013-0745-0.pdf]

   Grah, O., and J. Beaulieu, 2013: The effect of climate change on glacier ablation and baseflow support in the Nooksack River basin and implications on Pacific salmonid species protection and recovery. *Climatic Change*, **120**, 657-670, doi:10.1007/s10584-013-0747-y.

   McNutt, D., 2008: Native Peoples: The "Miners Canary" on Climate Change, 16 pp., Northwest Indian Applied Research Institute, Evergreen State College. [Available online at http://nwindian.evergreen.edu/pdf/climatechangereport.pdf]

7. Coastal Louisiana Tribal Communities, 2012: Stories of Change: Coastal Louisiana Tribal Communities' Experiences of a Transforming Environment (Grand Bayou, Grand Caillou/Dulac, Isle de Jean Charles, Pointe-au-Chien). Workshop Report Input Into the National Climate Assessment. Pointe-aux-Chenes, Louisiana.

8. Souza, K., and J. Tanimoto, 2012: PRiMO IKE Hui Technical Input for the National Climate Assessment – Tribal Chapter. PRiMO IKE Hui Meeting – January 2012, Hawai'i, 5 pp., U.S. Global Change Research Program, Washington, D.C.

9. BIA, cited 2012: Alaska Region Overview. U.S. Department of the Interior, Bureau of Indian Affairs. [Available online at http://www.bia.gov/WhoWeAre/RegionalOffices/Alaska/]

10. Brubaker, M. Y., J. N. Bell, J. E. Berner, and J. A. Warren, 2011: Climate change health assessment: A novel approach for Alaska Native communities. *International journal of circumpolar health*, **70**, doi:10.3402/ijch.v70i3.17820.

11. Hinzman, L. D., N. D. Bettez, W. R. Bolton, F. S. Chapin, III, M. B. Dyurgerov, C. L. Fastie, B. Griffith, R. D. Hollister, A. Hope, H. P. Huntington, A. M. Jensen, G. J. Jia, T. Jorgenson, D. L. Kane, D. R. Klein, G. Kofinas, A. H. Lynch, A. H. Lloyd, A. D. McGuire, F. E. Nelson, W. C. Oechel, T. E. Osterkamp, C. H. Racine, V. E. Romanovsky, R. S. Stone, D. A. Stow, M. Sturm, C. E. Tweedie, G. L. Vourlitis, M. D. Walker, D. A. Walker, P. J. Webber, J. M. Welker, K. S. Winker, and K. Yoshikawa, 2005: Evidence and

implications of recent climate change in Northern Alaska and other Arctic regions. *Climatic Change*, **72,** 251-298, doi:10.1007/s10584-005-5352-2. [Available online at http://www.springerlink.com/index/10.1007/s10584-005-5352-2]

Laidler, G. J., J. D. Ford, W. A. Gough, T. Ikummaq, A. S. Gagnon, S. Kowal, K. Qrunnut, and C. Irngaut, 2009: Travelling and hunting in a changing Arctic: Assessing Inuit vulnerability to sea ice change in Igloolik, Nunavut. *Climatic Change*, **94,** 363-397, doi:10.1007/s10584-008-9512-z.

Wang, M., and J. E. Overland, 2012: A sea ice free summer Arctic within 30 years: An update from CMIP5 models. *Geophysical Research Letters*, **39,** L18501, doi:10.1029/2012GL052868. [Available online at http://onlinelibrary.wiley.com/doi/10.1029/2012GL052868/pdf]

12. Pungowiyi, C., 2009: Siberian Yupic Elder, personal communication

    Pungowiyi, C., 2002: Special report on climate impacts in the Arctic. *Native Peoples-Native Homelands Climate Change Workshop: Final Report: Circles of Wisdom*, N. G. Maynard, Ed., NASA Goddard Space Flight Center, 11-12. [Available online at http://www.usgcrp.gov/usgcrp/Library/nationalassessment/native.pdf]

13. Parkinson, A. J., 2010: Sustainable development, climate change and human health in the Arctic. *Circumpolar Health*, **69,** 99-105. [Available online at http://www.circumpolarhealthjournal.net/index.php/ijch/article/view/17428]

Finding 10: Ecosystems
1. Millennium Ecosystem Assessment, 2005: *Ecosystems and Human Well-Being, Health Synthesis*. Island press 53 pp.

2. Staudt, A., A. K. Leidner, J. Howard, K. A. Brauman, J. S. Dukes, L. J. Hansen, C. Paukert, J. Sabo, and L. A. Solórzano, 2013: The added complications of climate change: Understanding and managing biodiversity and ecosystems. *Frontiers in Ecology and the Environment*, **11,** 494-501, doi:10.1890/120275. [Available online at http://www.esajournals.org/doi/pdf/10.1890/120275]

3. Woodbury, P. B., J. E. Smith, and L. S. Heath, 2007: Carbon sequestration in the US forest sector from 1990 to 2010. *Forest Ecology and Management*, **241,** 14-27, doi:10.1016/j.foreco.2006.12.008.

4. King, A. W., D. J. Hayes, D. N. Huntzinger, O. Tristram, T. O. West, and W. M. Post, 2012: North America carbon dioxide sources and sinks: Magnitude, attribution, and uncertainty. *Frontiers in Ecology and the Environment*, **10,** 512-519, doi:10.1890/120066.

5. Williams, C. A., G. J. Collatz, J. Masek, and S. N. Goward, 2012: Carbon consequences of forest disturbance and recovery across the conterminous United States. *Global Biogeochemical Cycles*, **26,** GB1005, doi:10.1029/2010gb003947.

6. Vose, J. M., D. L. Peterson, and T. Patel-Weynand, Eds., 2012: *Effects of Climatic Variability and Change on Forest Ecosystems: A Comprehensive Science Synthesis for the U.S. Forest Sector. General Technical Report PNW-GTR-870*. U.S. Department of Agriculture, Forest Service, Pacific Northwest Research Station, 265 pp. [Available online at http://www.nrsda.gov/oce/climate_change/effects_2012/FS_Climate1114%20opt.pdf]

7. Running, S. W., R. R. Nemani, F. A. Heinsch, M. Zhao, M. Reeves, and H. Hashimoto, 2004: A continuous satellite-derived measure of global terrestrial primary production. *BioScience*, **54,** 547-560, doi:10.1641/0006-3568(2004)054[0547:ACSMOG]2.0.CO;2.

[Available online at http://ecocast.arc.nasa.gov/pubs/pdfs/2004/Running_Bioscience.pdf]

8. Smith, W. B., P. D. Miles, C. H. Perry, and S. A. Pugh, 2009: Forest Resources of the United States, 2007. General Technical Report WO-78. 336 pp., U.S. Department of Agriculture. Forest Service, Washington, D.C. [Available online at http://www.fs.fed.us/nrs/pubs/gtr/gtr_wo78.pdf]

9. EPA, 2013: Annex 3.12. Methodology for estimating net carbon stock changes in forest land remaining forest lands. *Inventory of US Greenhouse Gas Emissions and Sinks: 1990-2011. Epa 430-R-13-001*, U.S. Environmental Protection Agency, A-254 - A-303. [Available online at http://www.epa.gov/climatechange/Downloads/ghgemissions/US-GHG-Inventory-2011-Annex_Complete_Report.pdf]

Woodall, C. W., K. Skog, J. E. Smith, and C. H. Perry, 2011: Maintenance of forest contribution to global carbon cycles (criterion 5). *National Report on Sustainable Forests -- 2010. FS-979*, G. Robertson, P. Gaulke, and R. McWilliams, Eds., U.S. Department of Agriculture, U.S. Forest Service, II-59 - II-65. [Available online at http://www.fs.fed.us/research/sustain/2010SustainabilityReport/documents/draft2010sustainabilityreport.pdf]

10. Allen, C. D., A. K. Macalady, H. Chenchouni, D. Bachelet, N. McDowell, M. Vennetier, T. Kitzberger, A. Rigling, D. D. Breshears, E. H. Hogg, P. Gonzalez, R. Fensham, Z. Zhang, J. Castro, N. Demidova, J.-H. Lim, G. Allard, S. W. Running, A. Semerci, and N. Cobb, 2010: A global overview of drought and heat-induced tree mortality reveals emerging climate change risks for forests. *Forest Ecology and Management*, **259,** 660-684, doi:10.1016/j.foreco.2009.09.001. [Available online at http://www.sciencedirect.com/science/article/pii/S037811270900615X]

Dukes, J. S., J. Pontius, D. Orwig, J. R. Garnas, V. L. Rodgers, N. Brazee, B. Cooke, K. A. Theoharides, E. E. Stange, R. Harrington, J. Ehrenfeld, J. Gurevitch, M. Lerdau, K. Stinson, R. Wick, and M. Ayres, 2009: Responses of insect pests, pathogens, and invasive plant species to climate change in the forests of northeastern North America: What can we predict? *Canadian Journal of Forest Research*, **39,** 231-248, doi:10.1139/X08-171. [Available online at http://www.nrcresearchpress.com/doi/pdf/10.1139/X08-171]

McDowell, N., W. T. Pockman, C. D. Allen, D. D. Breshears, N. Cobb, T. Kolb, J. Plaut, J. Sperry, A. West, E. A. Yepez, and D. G. Williams, 2008: Mechanisms of plant survival and mortality during drought: Why do some plants survive while others succumb to drought? *New Phytologist*, **178,** 719-739, doi:10.1111/j.1469-8137.2008.02436.x. [Available online at http://onlinelibrary.wiley.com/doi/10.1111/j.1469-8137.2008.02436.x/pdf]

11. Raffa, K. F., B. H. Aukema, B. J. Bentz, A. L. Carroll, J. A. Hicke, M. G. Turner, and W. H. Romme, 2008: Cross-scale drivers of natural disturbances prone to anthropogenic amplification: The dynamics of bark beetle eruptions. *BioScience*, **58,** 501-517, doi:10.1641/b580607. [Available online at http://www.jstor.org/stable/pdfplus/10.1641/B580607.pdf]

12. Van Mantgem, P. J., N. L. Stephenson, J. C. Byrne, L. D. Daniels, J. F. Franklin, P. Z. Fule, M. E. Harmon, A. J. Larson, J. M. Smith, A. H. Taylor, and T. T. Veblen, 2009: Widespread increase of tree mortality rates in the western United States. *Science*, **323,** 521-524, doi:10.1126/science.1165000.

13. Williams, A. P., C. D. Allen, C. I. Millar, T. W. Swetnam, J. Michaelsen, C. J. Still, and S. W. Leavitt, 2010: Forest responses to increasing aridity and warmth in the southwestern United States. *Proceedings of the National Academy of Sciences*, **107**, 21289-21294, doi:10.1073/pnas.0914211107. [Available online at http://www.pnas.org/content/107/50/21289.full]

14. Williams, A. P., C. D. Allen, A. K. Macalady, D. Griffin, C. A. Woodhouse, D. M. Meko, T. W. Swetnam, S. A. Rauscher, R. Seager, H. D. Grissino-Mayer, J. S. Dean, E. R.Cook, C. Gangodagamage, M. Cai, and N. G. McDowell, 2013: Temperature as a potent driver of regional forest drought stress and tree mortality. *Nature Climate Change*, **3**, 292-297, doi:10.1038/nclimate1693. [Available online at http://www.nature.com/nclimate/journal/v3/n3/pdf/nclimate1693.pdf]

15. Bowman, D. M. J. S., J. K. Balch, P. Artaxo, W. J. Bond, J. M. Carlson, M. A. Cochrane, C. M. D'Antonio, R. S. DeFries, J. C. Doyle, S. P. Harrison, F. H. Johnston, J. E. Keeley, M. A. Krawchuk, C. A. Kull, J. B. Marston, M. A. Moritz, I. C. Prentice, C. I. Roos, A. C. Scott, T. W. Swetnam, G. R. van der Werf, and S. J. Pyne, 2009: Fire in the Earth system. *Science*, **324**, 481-484, doi:10.1126/science.1163886.

Keane, R. E., J. K. Agee, P. Fulé, J. E. Keeley, C. Key, S. G. Kitchen, R. Miller, and L. A. Schulte, 2009: Ecological effects of large fires on US landscapes: Benefit or catastrophe? *International Journal of Wildland Fire*, **17**, 696-712, doi:10.1071/WF07148.

Littell, J. S., D. McKenzie, D. L. Peterson, and A. L. Westerling, 2009: Climate and wildfire area burned in western US ecoprovinces, 1916-2003. *Ecological Applications*, **19**, 1003-1021, doi:10.1890/07-1183.1.

NRC, 2011: *Climate Stabilization Targets: Emissions, Concentrations, and Impacts over Decades to Millennia*. National Research Council. The National Academies Press, 298 pp. [Available online at http://www.nap.edu/catalog.php?record_id=12877]

Westerling, A. L., M. G. Turner, E. A. H. Smithwick, W. H. Romme, and M. G. Ryan, 2011: Continued warming could transform Greater Yellowstone fire regimes by mid-21st century. *Proceedings of the National Academy of Sciences*, **108**, 13165-13170, doi:10.1073/pnas.1110199108. [Available online at http://www.pnas.org/content/early/2011/07/20/1110199108.abstract; http://www.pnas.org/content/108/32/13165.full.pdf]

16. Bierwagen, B. G., D. M. Theobald, C. R. Pyke, A. Choate, P. Groth, J. V. Thomas, and P. Morefield, 2010: National housing and impervious surface scenarios for integrated climate impact assessments. *Proceedings of the National Academy of Sciences*, **107**, 20887-20892, doi:10.1073/pnas.1002096107.

17. Radeloff, V. C., R. B. Hammer, S. I. Stewart, J. S. Fried, S. S. Holcomb, and J. F. McKeefry, 2005: The wildland-urban interface in the United States. *Ecological Applications*, **15**, 799-805, doi:10.1890/04-1413.

Theobald, D. M., and W. H. Romme, 2007: Expansion of the US wildland-urban interface. *Landscape and Urban Planning*, **83**, 340-354, doi:10.1016/j.landurbplan.2007.06.002.

18. Stephens, S. L., M. A. Adams, J. Handmer, F. R. Kearns, B. Leicester, J. Leonard, and M. A. Moritz, 2009: Urban–wildland fires: How California and other regions of the US can learn from Australia. *Environmental Research Letters*, **4**, 014010, doi:10.1088/1748-9326/4/1/014010.

19. Westerling, A. L., H. G. Hidalgo, D. R. Cayan, and T. W. Swetnam, 2006: Warming and earlier spring increase western U.S. forest wildfire activity. *Science*, **313**, 940-943, doi:10.1126/science.1128834.

20. Galloway, J. N., A. R. Townsend, J. W. Erisman, M. Bekunda, Z. C. Cai, J. R. Freney, L. A. Martinelli, S. P. Seitzinger, and M. A. Sutton, 2008: Transformation of the nitrogen cycle: Recent trends, questions, and potential solutions. *Science*, **320**, 889-892, doi:10.1126/science.1136674.

Vitousek, P. M., J. D. Aber, R. W. Howarth, G. E. Likens, P. A. Matson, D. W. Schindler, W. H. Schlesinger, and D. G. Tilman, 1997: Human alteration of the global nitrogen cycle: Sources and consequences. *Ecological Applications*, **7**, 737-750, doi:10.1890/1051-0761(1997)007[0737:HAOTGN]2.0.CO;2.

21. Harley, C. D. G., 2011: Climate change, keystone predation, and biodiversity loss. *Science*, **334**, 1124-1127, doi:10.1126/science.1210199.

22. Rode, K. D., S. C. Amstrup, and E. V. Regehr, 2010: Reduced body size and cub recruitment in polar bears associated with sea ice decline. *Ecological Applications*, **20**, 768-782, doi:10.1890/08-1036.1.

23. Moritz, C., J. L. Patton, C. J. Conroy, J. L. Parra, G. C. White, and S. R. Beissinger, 2008: Impact of a century of climate change on small-mammal communities in Yosemite National Park, USA. *Science*, **322**, 261-264, doi:10.1126/science.1163428.

24. Anderegg, W. R. L., J. M. Kane, and L. D. L. Anderegg, 2012: Consequences of widespread tree mortality triggered by drought and temperature stress. *Nature Climate Change*, **3**, 30-36, doi:10.1038/nclimate1635.

25. Post, E., C. Pedersen, C. C. Wilmers, and M. C. Forchhammer, 2008: Warming, plant phenology and the spatial dimension of trophic mismatch for large herbivores. *Proceedings of the Royal Society B: Biological Sciences*, **275**, 2005-2013, doi:10.1098/rspb.2008.0463. [Available online at http://rspb.royalsocietypublishing.org/content/275/1646/2005.full.pdf+html]

26. Crausbay, S. D., and S. C. Hotchkiss, 2010: Strong relationships between vegetation and two perpendicular climate gradients high on a tropical mountain in Hawai'i. *Journal of Biogeography*, **37**, 1160-1174, doi:10.1111/j.1365-2699.2010.02277.x.

27. Garroway, C. J., J. Bowman, T. J. Cascaden, G. L. Holloway, C. G. Mahan, J. R. Malcolm, M. A. Steele, G. Turner, and P. J. Wilson, 2009: Climate change induced hybridization in flying squirrels. *Global Change Biology*, **16**, 113-121, doi:10.1111/j.1365-2486.2009.01948.x. [Available online at http://onlinelibrary.wiley.com/doi/10.1111/j.1365-2486.2009.01948.x/pdf]

28. Dunnell, K. L., and S. E. Travers, 2011: Shifts in the flowering phenology of the Northern Great Plains: Patterns over 100 years *American Journal of Botany*, **98**, 935-945, doi:10.3732/ajb.1000363. [Available online at http://www.amjbot.org/content/98/6/935.full.pdf+html]

29. Nye, J. A., J. S. Link, J. A. Hare, and W. J. Overholtz, 2009: Changing spatial distribution of fish stocks in relation to climate and population size on the Northeast United States continental shelf. *Marine Ecology Progress Series*, **393**, 111-129, doi:10.3354/meps08220.

30. Sperry, J. H., G. Blouin-Demers, G. L. F. Carfagno, and P. J. Weatherhead, 2010: Latitudinal variation in seasonal activity and mortality in ratsnakes (*Elaphe obsoleta*). *Ecology*, **91**, 1860-1866, doi:10.1890/09-1154.1.

31. Van Buskirk, J., R. S. Mulvihill, and R. C. Leberman, 2008: Variable shifts in spring and autumn migration phenology in North American songbirds associated with climate change. *Global Change Biology*, **15**, 760-771, doi:10.1111/j.1365-2486.2008.01751.x.

32. Botero, C. A., and D. R. Rubenstein, 2012: Fluctuating environments, sexual selection and the evolution of flexible mate choice in birds. *PLoS ONE*, **7**, e32311, doi:10.1371/journal.pone.0032311.

33. Ibáñez, I., J. S. Clark, and M. C. Dietze, 2008: Evaluating the sources of potential migrant species: Implications under climate change. *Ecological Applications*, **18**, 1664-1678, doi:10.1890/07-1594.1.

34. Fodrie, F., K. L. Heck, S. P. Powers, W. M. Graham, and K. L. Robinson, 2009: Climate-related, decadal-scale assemblage changes of seagrass-associated fishes in the northern Gulf of Mexico. *Global Change Biology*, **16**, 48-59, doi:10.1111/j.1365-2486.2009.01889.x. [Available online at http://onlinelibrary.wiley.com/doi/10.1111/j.1365-2486.2009.01889.x/pdf]

35. Wood, A. J. M., J. S. Collie, and J. A. Hare, 2009: A comparison between warm-water fish assemblages of Narragansett Bay and those of Long Island Sound waters. *Fishery Bulletin*, **107**, 89-100.

Finding 11: Oceans

1. NMFS, 2012: Fisheries of the United States 2011, 139 pp., National Marine Fisheries Service, Office of Science and Technology, Silver Spring, MD. [Available online at http://www.st.nmfs.noaa.gov/st1/fus/fus11/FUS_2011.pdf]

    NOC, 2012: National Ocean Policy Draft Implementation Plan, 118 pp., National Ocean Council, Washington, D.C. [Available online at http://www.whitehouse.gov/sites/default/files/microsites/ceq/national_ocean_policy_draft_implementation_plan_01-12-12.pdf]

    U.S. Commission on Ocean Policy, 2004: An Ocean Blueprint for the 21st Century: Final Report, 28 pp., U.S. Commission on Ocean Policy, Washington, D.C. [Available online at http://govinfo.library.unt.edu/oceancommission/documents/full_color_rpt/000_ocean_full_report.pdf]

2. NRC, 2010: Ocean Acidification. A National Strategy to Meet the Challenges of a Changing Ocean, 175 pp., Committee on the Development of an Integrated Science Strategy for Ocean Acidification Monitoring Research and Impacts Assessment, Ocean Studies Board, Division on Earth and Life Studies, National Research Council, Washington, D.C. [Available online at http://www.nap.edu/catalog.php?record_id=12904]

3. Chavez, F. P., M. Messié, and J. T. Pennington, 2011: Marine primary production in relation to climate variability and change. *Annual Review of Marine Science*, **3**, 227-260, doi:10.1146/annurev.marine.010908.163917.

4. Etheridge, D. M., et al., 2010: Law Dome Ice Core 2000-Year CO$_2$, CH$_4$, and N$_2$O Data, IGBP PAGES/World Data Center for Paleoclimatology. Data Contribution Series #2010-070. NOAA/NCDC Paleoclimatology Program, Boulder, CO.

NCDC, cited 2012: Extended Reconstructed Sea Surface Temperature NOAA's National Climatic Data Center. [Available online at http://www.ncdc.noaa.gov/ersst/]

Smith, T. M., R. W. Reynolds, T. C. Peterson, and J. Lawrimore, 2008: Improvements to NOAA's historical merged land-ocean surface temperature analysis (1880-2006). *Journal of Climate*, **21**, 2283-2296, doi:10.1175/2007JCLI2100.1.

CSIRO, cited 2012: The Commonwealth Scientific and Industrial Research Organisation [Available online at www.csiro.au/]

Church, J. A., and N. J. White, 2011: Sea-level rise from the late 19th to the early 21st century. *Surveys in Geophysics*, **32**, 585-602, doi:10.1007/s10712-011-9119-1.

University of Illinois, 2012: Sea Ice Dataset, University of Illinois, Department of Atmospheric Sciences, Urbana, IL. [Available online at http://arctic.atmos.uiuc.edu/SEAICE/]

Doney, S. C., M. Ruckelshaus, J. E. Duffy, J. P. Barry, F. Chan, C. A. English, H. M. Galindo, J. M. Grebmeier, A. B. Hollowed, N. Knowlton, J. Polovina, N. N. Rabalais, W. J. Sydeman, and L. D. Talley, 2012: Climate change impacts on marine ecosystems. *Annual Review of Marine Science*, **4**, 11-37, doi:10.1146/annurev-marine-041911-111611. [Available online at http://www.annualreviews.org/eprint/fzUZd7Z748TeHmB7p8cn/full/10.1146/annurev-marine-041911-111611]

5. Tans, P., and R. Keeling, cited 2012: Trends in Atmospheric Carbon Dioxide, Full Mauna Loa CO$_2$ Record. NOAA's Earth System Research Laboratory. [Available online at http://www.esrl.noaa.gov/gmd/ccgg/trends/]

6. MacFarling Meure, C., D. Etheridge, C. Trudinger, P. Steele, R. Langenfelds, and T. van Ommen, 2006: Law Dome CO$_2$, CH$_4$, and N$_2$O ice core records extended to 2000 years BP. *Geophysical Research Letters*, **33**, L14810, doi:10.1029/2006GL026152. [Available online at http://onlinelibrary.wiley.com/doi/10.1029/2006GL026152/pdf]

7. Sabine, C. L., R. A. Feely, N. Gruber, R. M. Key, K. Lee, J. L. Bullister, R. Wanninkhof, C. S. Wong, D. W. R. Wallace, B. Tilbrook, F. J. Millero, T.-H. Peng, A. Kozyr, T. Ono, and A. F. Rios, 2004: The oceanic sink for anthropogenic CO$_2$. *Science*, **305**, 367-371, doi:10.1126/science.1097403.

8. Feely, R. A., S. C. Doney, and S. R. Cooley, 2009: Ocean acidification: Present conditions and future changes in a high-CO$_2$ world. *Oceanography*, **22**, 36-47, doi:10.5670/oceanog.2009.95. [Available online at http://www.tos.org/oceanography/archive/22-4_feely.pdf]

9. Feely, R. A., C. L. Sabine, J. M. Hernandez-Ayon, D. Ianson, and B. Hales, 2008: Evidence for upwelling of corrosive "acidified" water onto the continental shelf. *Science*, **320**, 1490-1492, doi:10.1126/science.1155676. [Available online at http://www.sciencemag.org/content/320/5882/1490.short]

10. Tribollet, A., C. Godinot, M. Atkinson, and C. Langdon, 2009: Effects of elevated pCO$_2$ on dissolution of coral carbonates by microbial euendoliths. *Global Biogeochemical Cycles*, **23**, GB3008, doi:10.1029/2008GB003286.

Wisshak, M., C. H. L. Schönberg, A. Form, and A. Freiwald, 2012: Ocean acidification accelerates reef bioerosion. *PLoS ONE*, **7**, e45124, doi:10.1371/journal.pone.0045124. [Available online at http://www.plosone.org/article/fetchObject.action?uri=info%3Adoi%2F10.1371%2Fjournal.pone.0045124&representation=PDF]

11. Cooley, S. R., H. L. Kite-Powell, L. Hauke, and S. C. Doney, 2009: Ocean acidification's potential to alter global marine ecosystem services. *Oceanography*, **22**, 172-181, doi:10.5670/oceanog.2009.106.

Doney, S. C., W. M. Balch, V. J. Fabry, and R. A. Feely, 2009: Ocean acidification: A critical emerging problem for the ocean sciences. *Oceanography*, **22**, 16-25, doi:10.5670/oceanog.2009.93. [Available online at https://darchive.mblwhoilibrary.org/bitstream/handle/1912/3181/22-4_doney.pdf?sequence=1]

Kroeker, K. J., R. L. Kordas, R. Crim, I. E. Hendriks, L. Ramajo, G. S. Singh, C. M. Duarte, and J.-P. Gattuso, 2013: Impacts of ocean acidification on marine organisms: Quantifying sensitivities and interaction with warming. *Global Change Biology*, **19**, 1884-1896, doi:10.1111/gcb.12179. [Available online at http://onlinelibrary.wiley.com/doi/10.1111/gcb.12179/pdf]

Kroeker, K. J., R. L. Kordas, R. N. Crim, and G. G. Singh, 2010: Meta-analysis reveals negative yet variable effects of ocean acidification on marine organisms. *Ecology Letters*, **13**, 1419-1434, doi:10.1111/j.1461-0248.2010.01518.x. [Available online at http://onlinelibrary.wiley.com/doi/10.1111/j.1461-0248.2010.01518.x/pdf]

12. Marshall, P., and H. Schuttenberg, 2006: *A Reef Manager's Guide to Coral Bleaching*. Great Barrier Reef Marine Park Authority, IUCN Global Marine Programme, and U.S. National Oceanic and Atmospheric Administration, 163 pp. [Available online at http://data.iucn.org/dbtw-wpd/edocs/2006-043.pdf]

13. Talmage, S. C., and C. J. Gobler, 2010: Effects of past, present, and future ocean carbon dioxide concentrations on the growth and survival of larval shellfish. *Proceedings of the National Academy of Sciences*, **107**, 17246-17251, doi:10.1073/pnas.0913804107. [Available online at http://www.sciencedaily.com/releases/2010/09/100928154754.htm]

14. Bates, A. E., W. B. Stickle, and C. D. G. Harley, 2010: Impact of temperature on an emerging parasitic association between a sperm-feeding scuticociliate and Northeast Pacific sea stars. *Journal of Experimental Marine Biology and Ecology*, **384**, 44-50, doi:10.1016/j.jembe.2009.12.001.

Eakin, C. M., J. A. Morgan, S. F. Heron, T. B. Smith, G. Liu, L. Alvarez-Filip, B. Baca, E. Bartels, C. Bastidas, C. Bouchon, M. Brandt, A. W. Bruckner, L. Bunkley-WIlliams, A. Cameron, B. D. Causey, M. Chiappone, T. R. L. Christensen, M. J. C. Crabbe, O. Day, E. de la Guardia, G. Díaz-Pulido, D. Di Resta, D. L. Gil-Agudelo, D. S. Gilliam, R. N. Ginsburg, S. Gore, H. M. Guzmán, J. C. Hendee, E. A. Hernández-Delgado, E. Husain, C. F. G. Jeffrey, R. J. Jones, E. Jordán-Dahlgren, L. S. Kaufman, D. I. Kline, P. A. Kramer, J. C. Lang, D. Lirman, J. Mallela, C. Manfrino, J.-P. Maréchal, K. Marks, J. Mihaly, W. J. Miller, E. M. Mueller, E. M. Muller, C. A. Orozco Toro, H. A. Oxenford, D. Ponce-Taylor, N. Quinn, K. B. Ritchie, S. Rodríguez, A. Rodríguez Ramírez, S. Romano, J. F. Samhouri, J. A. Sánchez, G. P. Schmahl, B. V. Shank, W. J. Skirving, S. C. C. Steiner, E. Villamizar, S. M. Walsh, C. Walter, E. Weil, E. H. Williams, K. W. Roberson, and Y. Y., 2010: Caribbean corals in crisis: Record thermal stress, bleaching, and mortality in 2005. *PLoS ONE*, **5**, e13969, doi:10.1371/journal.

pone.0013969. [Available online at http://www.plosone.org/article/info%3Adoi%2F10.1371%2Fjournal.pone.0013969]

Harvell, D., S. Altizer, I. M. Cattadori, L. Harrington, and E. Weil, 2009: Climate change and wildlife diseases: When does the host matter the most? *Ecology*, **90**, 912-920, doi:/10.1890/08-0616.1.

Staehli, A., R. Schaerer, K. Hoelzle, and G. Ribi, 2009: Temperature induced disease in the starfish *Astropecten jonstoni*. *Marine Biodiversity Records*, **2**, e78, doi:10.1017/S1755267209000633. [Available online at http://journals.cambridge.org/action/displayAbstract?fromPage=online&aid=5466240]

Ward, J. R., and K. D. Lafferty, 2004: The elusive baseline of marine disease: Are diseases in ocean ecosystems increasing? *PLoS Biology*, **2**, e120, doi:10.1371/journal.pbio.0020120.

15. Bruno, J. F., E. R. Selig, K. S. Casey, C. A. Page, B. L. Willis, C. D. Harvell, H. Sweatman, and A. M. Melendy, 2007: Thermal stress and coral cover as drivers of coral disease outbreaks. *PLoS Biology*, **5**, e124, doi:10.1371/journal.pbio.0050124.

16. Boyett, H. V., D. G. Bourne, and B. L. Willis, 2007: Elevated temperature and light enhance progression and spread of black band disease on staghorn corals of the Great Barrier Reef. *Marine Biology*, **151**, 1711-1720, doi:10.1007/200227-006-0603-y.

Ward, J. R., K. Kim, and C. D. Harvell, 2007: Temperature affects coral disease resistance and pathogen growth. *Marine Ecology Progress Series*, **329**, 115-121, doi:10.3354/meps329115.

Case, R. J., S. R. Longford, A. H. Campbell, A. Low, N. Tujula, P. D. Steinberg, and S. Kjelleberg, 2011: Temperature induced bacterial virulence and bleaching disease in a chemically defended marine macroalga. *Environmental Microbiology*, **13**, 529-537, doi:10.1111/j.1462-2920.02356.x.

17. Hughes, J. E., L. A. Deegan, J. C. Wyda, M. J. Weaver, and A. Wright, 2002: The effects of eelgrass habitat loss on estuarine fish communities of southern New England. *Estuaries and Coasts*, **25**, 235-249, doi:10.1007/BF02691311.

18. Bjork, M., F. Short, E. McLeod, and S. Beer, 2008: *Managing Seagrasses for Resilience to Climate Change*. World Conservation Union.

19. Pinsky, M. L., and M. Fogarty, 2012: Lagged social-ecological responses to climate and range shifts in fisheries. *Climatic Change*, **115**, 883-891, doi:10.1007/s10584-012-0599-x.

Finding 12: Responses

1. Bierbaum, R. M., D. G. Brown, and J. L. McAlpine, 2008: *Coping with Climate Change: National Summit Proceedings*. University of Michigan Press, 256 pp.

2. SEGCC, 2007: Confronting Climate Change: Avoiding the Unmanageable and Managing the Unavoidable. Report Prepared for the United Nations Commission on Sustainable Development. R. Bierbaum, J. P. Holdren, M. MacCracken, R. H. Moss, P. H. Raven, and H. J. Schellnhuber, Eds., 144 pp., Scientific Expert Group on Climate Change, Sigma Xi and the United Nations Foundation, Research Triangle Park, NC and Washington, D.C. . [Available online at http://www.globalproblems-globalsolutions-files.org/unf_website/PDF/climate%20change_avoid_unmanagable_manage_unavoidable.pdf]

3. McMullen, C. P., and J. R. Jabbour, 2009: *Climate Change Science Compendium 2009*. United Nations Environment Programme.

    Skaggs, R., T. C. Janetos, K. A. Hibbard, and J. S. Rice, 2012: Climate and Energy-Water-Land System Interactions Technical Report to the U.S. Department of Energy in Support of the National Climate Assessment, 152 pp., Pacific Northwest National Laboratory, Richland, Washington. [Available online at http://climatemodeling. science.energy.gov/f/PNNL-21185_FINAL_REPORT.pdf]

    Wilbanks, T., D. Bilello, D. Schmalzer, and M. Scott, 2012: Climate Change and Energy Supply and Use. Technical Report to the U.S. Department of Energy in Support of the National Climate Assessment, 79 pp., Oak Ridge National Laboratory, U.S. Department of Energy, Office of Science, Oak Ridge, TN. [Available online at http://www.esd.ornl.gov/eess/EnergySupplyUse.pdf]

    Wilbanks, T., S. Fernandez, G. Backus, P. Garcia, K. Jonietz, P. Kirshen, M. Savonis, B. Solecki, and L. Toole, 2012: Climate Change and Infrastructure, Urban Systems, and Vulnerabilities. Technical Report to the U.S. Department of Energy in Support of the National Climate Assessment, 119 pp., Oak Ridge National Laboratory. U.S. Department of Energy, Office of Science, Oak Ridge, TN. [Available online at http://www.esd.ornl.gov/eess/ Infrastructure.pdf]

4. Karl, T. R., J. T. Melillo, and T. C. Peterson, Eds., 2009: *Global Climate Change Impacts in the United States*. Cambridge University Press, 189 pp. [Available online at http://downloads.globalchange.gov/ usimpacts/pdfs/climate-impacts-report.pdf]

5. NRC, 2010: *Adapting to Impacts of Climate Change. America's Climate Choices: Report of the Panel on Adapting to the Impacts of Climate Change*. National Research Council. The National Academies Press, 292 pp. [Available online at http://www.nap.edu/catalog.php?record_ id=12783]

6. The White House, 2013: Executive Order 13653. Preparing the United States for the Impacts of Climate Change. The White House, Washington, D.C. [Available online at http://www.whitehouse.gov/ the-press-office/2013/11/01/executive-order-preparing-united- states-impacts-climate-change]

7. Goulder, L. H., and R. N. Stavins, 2011. Challenges from state-federal interactions in US climate change policy. *The American Economic Review*, **101,** 253-257, doi:10.1257/aer.101.3.253.

    Morsch, A., and R. Bartlett, 2011: Policy Brief: State Strategies to Plan for and Adapt to Climate Change - NI PB 11-08, 11 pp., Nicholas Institute for Environmental Policy Solutions – Duke University, Durham, NC. [Available online at http://nicholasinstitute.duke. edu/sites/default/files/publications/state-strategies-to-plan-for- and-adapt-to-climate-change-paper.pdf]

8. Feldman, I. R., and J. H. Kahan, 2007: Preparing for the day after tomorrow: Frameworks for climate change adaptation. *Sustainable Development Law & Policy*, **8,** 31-39, 87-89. [Available online at http://digitalcommons.wcl.american.edu/cgi/viewcontent. cgi?article=1162&context=sdlp]

    Moser, S. C., 2009: Good Morning America! The Explosive Awakening of the US to Adaptation, 39 pp., California Energy Commission, NOAA-Coastal Services Center, Sacramento, CA and Charleston, SC. [Available online at http://www.preventionweb. net/files/11374_MoserGoodMorningAmericaAdaptationin.pdf]

9. C2ES, cited 2013: State and Local Climate Adaptation. Center for Climate and Energy Solutions. [Available online at http://www. c2es.org/us-states-regions/policy-maps/adaptation]

10. Simmonds, J., 2011: Resource for Consideration by the NCA Teams Addressing the Impacts of Climate Change on Native Communities. Native Communities and Climate Change Project of the University of Colorado Law School and the Cooperative Institute for Research in Environmental Science.

11. Lamb, R., and M. V. Davis, 2011: Promoting Generations of Self Reliance: Stories and Examples of Tribal Adaptation to Change, 27 pp., U.S. Environmental Protection Agency Region 10, Seattle, WA. [Available online at http://www.epa.gov/region10/pdf/tribal/ stories_and_examples_of_tribal_adaptation_to_change.pdf]

12. Carmin, J., N. Nadkarni, and C. Rhie, 2012: Progress and Challenges in Urban Climate Adaptation Planning: Results of a Global Survey, 30 pp., Massachusetts Institute of Technology, ICLEI Local Governments for Sustainability, Cambridge, MA. [Available online at http://web.mit.edu/jcarmin/www/urbanadapt/Urban%20 Adaptation%20Report%20FINAL.pdf]

13. Dierwechter, Y., 2010: Metropolitan geographies of US climate action: Cities, suburbs, and the local divide in global responsibilities. *Journal of Environmental Policy & Planning*, **12,** 59-82, doi:10.1080/1523 9081003625960.

    Grannis, J., 2011: Adaptation Tool Kit: Sea Level Rise and Coastal Land Use. How Governments Can Use Land-Use Practices to Adapt to Sea-Level Rise, 100 pp., Georgetown Climate Center, Washington, D.C. [Available online at http://www.georgetownclimate.org/sites/ default/files/Adaptation_Tool_Kit_SLR.pdf]

    Kahn, M. E., 2009: Urban growth and climate change. *Annual Review of Resource Economics*, **1,** 333-350, doi:10.1146/annurev. resource.050708.144249.

    Selin, H., and S. D. VanDeveer, 2007: Political science and prediction: What's next for U.S. climate change policy? *Review of Policy Research*, **24,** 1-27, doi:10.1111/j.1541-1338.2007.00265.x. [Available online at http://pubpages.unh.edu/~sdv/US Climate-Policy.pdf]

    Solecki, W., and C. Rosenzweig, Eds., 2012: U.S. Cities and Climate Change: Urban, Infrastructure, and Vulnerability Issues, Technical Input Report Series, U.S. National Climate Assessment. U.S. Global Change Research Program, Washington, D.C.

    Tang, Z., S. D. Brody, C. Quinn, L. Chang, and T. Wei, 2010: Moving from agenda to action: Evaluating local climate change action plans. *Journal of Environmental Planning and Management*, **53,** 41- 62, doi:10.1080/09640560903399772.

14. Staudinger, M. D., N. B. Grimm, A. Staudt, S. L. Carter, F. S. Chapin, III, P. Kareiva, M. Ruckelshaus, and B. A. Stein, 2012: Impacts of Climate Change on Biodiversity, Ecosystems, and Ecosystem Services. Technical Input to the 2013 National Climate Assessment 296 pp., U.S. Geological Survey, Reston, VA. [Available online at http://downloads.usgcrp.gov/NCA/Activities/Biodiversity- Ecosystems and-Ecosystem-Services-Technical-Input.pdf]

REFERENCES

Colson, M., K. Heery, and A. Wallis, 2011: A Survey Of Regional Planning For Climate Adaptation, 20 pp., The National Association of Regional Councils, Washington, DC. [Available online at http://narc.org/wp-content/uploads/NOAA_White_Paper-FINAL2.pdf]

15. CDP, 2011: CDP S&P 500 Report: Strategic Advantage Through Climate Change Action, 49 pp., Carbon Disclosure Project, New York, NY and London, UK. [Available online at https://www.cdproject.net/CDPResults/CDP-2011-SP500.pdf]

16. Agrawala, S., M. Carraro, N. Kingsmill, E. Lanzi, M. Mullan, and G. Prudent-Richard, 2011: Private sector engagement in adaptation to climate change: Approaches to managing climate risks. *OECD Environment Working Papers*, **39**, doi:10.1787/5kg221jkf1g7-en.

Dell, J., and P. Pasteris, 2010: Adaptation in the Oil and Gas Industry to Projected Impacts of Climate Change. Society of Petroleum Engineers, 16 pp.

Oxfam America, cited 2012: The New Adaptation Marketplace: Climate Change and Opportunities for Green Economic Growth. Oxfam America. [Available online at http://www.usclimatenetwork.org/resource-database/the-new-adaptation-marketplace.pdf]

PWC, 2010: Business Leadership on Climate Change Adaptation: Encouraging Engagement and Action, 36 pp., PricewaterhouseCoopers LLP London, UK. [Available online at http://www.ukmediacentre.pwc.com/imagelibrary/downloadMedia.ashx?MediaDetailsID=1837]

17. Bierbaum, R., J. B. Smith, A. Lee, L. Carter, F. S. Chapin, III, P. Fleming, S. Ruffo, S. McNeeley, M. Stults, E. Wasley, and L. Verduzco, 2013 A comprehensive review of climate adaptation in the United States: More than before, but less than needed. *Mitigation and Adaptation Strategies for Global Change*, **18**, 361-406, doi:10.1007/s11027-012-9423-1. [Available online at http://link.springer.com/article/10.1007%2Fs11027-012-9423-1]

18. SFRCCC, 2012: A Region Responds to a Changing Climate. Southeast Florida Regional Climate Change Compact Counties. Regional Climate Action Plan, 80 pp., South Florida Regional Climate Change Compact Broward, Miami-Dade, Monroe, and Palm Beach Counties, FL. [Available online at http://southeastfloridaclimatecompact.org/pdf/Regional%20Climate%20Action%20Plan%20FINAL%20ADA%20Compliant.pdf]

19. EPA, 2013: Inventory of US Greenhouse Gas Emissions and Sinks: 1990-2011. U.S. Environmental Protection Agency, Washington, D.C. [Available online at http://www.epa.gov/climatechange/Downloads/ghgemissions/US-GHG-Inventory-2013-Main-Text.pdf]

20. The White House, 2010: Economic Report of the President, Council of Economic Advisors, 462 pp., The White House, Washington, D.C. [Available online at http://www.whitehouse.gov/sites/default/files/microsites/economic-report-president.pdf]

——, 2010: *Federal Climate Change Expenditures: Report to Congress*. Office of Management and Budget, 34 pp.

——, 2012: A Secure Energy Future: Progress Report. [Available online at http://www.whitehouse.gov/sites/default/files/email-files/the_blueprint_for_a_secure_energy_future_oneyear_progress_report.pdf]

DOE, 2009: Strategies of the Commercialization and Deployment of Greenhouse Gas Intensity-Reducing Technologies and Practices. DOE/PI-000, 190 pp., the Committee on Climate Change Science and Technology Integration [Available online at http://www.climatetechnology.gov/Strategy-Intensity-Reducing-Technologies.pdf]

GAO, 2011: Climate Change: Improvements Needed to Clarify National Priorities and Better Align Them with Federal Funding Decisions. GAO-11-317, 95 pp., U.S. Government Accountability Office. [Available online at http://www.gao.gov/assets/320/318556.pdf]

21. The White House, cited 2013: The President's Climate Action Plan. The White House. [Available online at http://www.whitehouse.gov/share/climate-action-plan]

22. Janetos, A., and A. Wagener, 2002: Understanding the Ancillary Effects of Climate Change Policies: A Research Agenda. World Resources Institute Policy Brief, Washington, D.C. [Available online at http://pdf.wri.org/climate_janetos_ancillary.pdf]

Haines, A., K. R. Smith, D. Anderson, P. R. Epstein, A. J. McMichael, I. Roberts, P. Wilkinson, J. Woodcock, and J. Woods, 2007: Policies for accelerating access to clean energy, improving health, advancing development, and mitigating climate change. *The Lancet*, **370**, 1264-1281, doi:10.1016/S0140-6736(07)61257-4.

23. Nemet, G. F., T. Holloway, and P. Meier, 2010: Implications of incorporating air-quality co-benefits into climate change policymaking. *Environmental Research Letters*, **5**, 014007, doi:10.1088/1748-9326/5/1/014007. [Available online at http://iopscience.iop.org/1748-9326/5/1/014007/pdf/1748-9326_5_1_014007.pdf]

24. Burtraw, D., A. Krupnick, K. Palmer, A. Paul, M. Toman, and C. Bloyd, 2003: Ancillary benefits of reduced air pollution in the US from moderate greenhouse gas mitigation policies in the electricity sector. *Journal of Environmental Economics and Management*, **45**, 650-673, doi:10.1016/S0095-0696(02)00022-0.

25. West, J. J., A. M. Fiore, L. W. Horowitz, and D. L. Mauzerall, 2006: Global Health Benefits of Mitigating ozone pollution with methane emission controls. *Proceedings of the National Academy of Sciences*, **103**, 3998-3993, doi:10.1073/pnas.0600201103. [Available online at http://www.pnas.org/content/103/11/3988.full.pdf+html]

26. NRC, 2010: *Limiting the Magnitude of Future Climate Change. America's Climate Choices. Panel on Limiting the Magnitude of Future Climate Change.* National Research Council, Board on Atmospheric Sciences and Climate, Division of Earth and Life Studies. The National Academies Press, 276 pp. [Available online at http://www.nap.edu/catalog.php?record_id=12785]

27. U.S. Mayors Climate Protection Agreement, cited 2012: List of Participating Mayors. U.S. Mayors Climate Protection Center, The U.S. Conference of Mayors. [Available online at http://www.usmayors.org/climateprotection/list.asp]

28. Beratan, K. K., and H. A. Karl, 2012: Ch. 10: Managing the science-policy interface in a complex and contentious world. *Restoring Lands - Coordinating Science, Politics and Action: Complexities of Climate and Governance*, H. A. Karl, L. Scarlett, J. C. Vargas-Moreno, and M. Flaxman, Eds., Springer, 183-216.

Mattson, D., H. Karl, and S. Clark, 2012: Ch. 12: Values in natural resource management and policy. *Restoring Lands - Coordinating Science, Politics and Action: Complexities of Climate and Governance*, H. A. Karl, L. Scarlett, J. C. Vargas-Moreno, and M. Flaxman, Eds., Springer, 239-259.

29. NRC, 2009: Informing Decisions in a Changing Climate. National Research Council, Panel on Strategies and Methods for Climate-Related Decision Support, Committee on the Human Dimensions of Global Change, Division of Behavioral and Social Sciences and Education. National Academies Press, 200 pp. [Available online at http://www.nap.edu/catalog.php?record_id=12626]

30. Lee, K. N., 1993: *Compass and Gyroscope: Integrating Science and Politics for the Environment*. Island Press, 255 pp.

31. NRC, 2010: Informing an Effective Response to Climate Change. America's Climate Choices: Panel on Informing Effective Decisions and Actions Related to Climate Change. National Research Council, Board on Atmospheric Sciences and Climate, Division on Earth and Life Studies, National Academies Press, 348 pp. [Available online at http://www.nap.edu/catalog.php?record_id=12784]

Willows, R. I., and R. K. Connell, Eds., 2003: Climate Adaptation: Risk, Uncertainty and Decision-Making. UKCIP Technical Report. UK Climate Impacts Programme. 166 pp. [Available online at http://www.ukcip.org.uk/wordpress/wp-content/PDFs/UKCIP-Risk-framework.pdf]

32. Kahan, D. M., and D. Braman, 2006: Cultural cognition and public policy. Yale Law & Policy Review, 24, 149-172.

33. Hall, J. W., R. J. Lempert, K. Keller, A. Hackbarth, C. Mijere, and D. J. McInerney, 2012: Robust climate policies under uncertainty: A comparison of robust decision making and info-gap methods. Risk Analysis, 32, 1657-1672, doi:10.1111/j.1539-6924.2012.01802.x.

Lempert, R. J., S. W. Popper, and S. C. Bankes, 2003: Shaping the Next One Hundred Years: New Methods for Quantitative, Long-Term Policy Analysis. Rand Corporation, 186 pp. [Available online at http://www.rand.org/pubs/monograph_reports/2007/MR1626.pdf]

Northeast

1. Groisman, P. Y., R. W. Knight, and O. G. Zolina, 2013: Recent trends in regional and global intense precipitation patterns. *Climate Vulnerability*, R. A. Pielke, Sr., Ed., Academic Press, 25-55.

2. Blake, E. S., T. B. Kimberlain, R. J. Berg, J. P. Cangialosi, and J. L. Beven, II 2013: Tropical Cyclone Report: Hurricane Sandy. (AL182012) 22 – 29 October 2012, 157 pp., National Oceanic and Atmospheric Administration, National Hurricane Center [Available online at http://www.nhc.noaa.gov/data/tcr/AL182012_Sandy.pdf]

3. NOAA, cited 2013: Billion Dollar Weather/Climate Disasters, List of Events. National Oceanic and Atmospheric Administration [Available online at http://www.ncdc.noaa.gov/billions/events]

4. PWD, cited 2013: Green City, Clean Waters. Philadelphia Water Department. [Available online at http://www.phillywatersheds.org/ltcpu/]

5. Jain, S., E. Stancioff, and A. Gray, cited 2012: Coastal Climate Adaptation in Maine's Coastal Communities: Governance Mapping for Culvert Management. [Available online at http://umaine.edu/maineclimatenews/archives/spring-2012/coastal-climate-adaptation in-maines-coastal-communities/]

6. NOAA, cited 2013: Sea Level Trends. National Oceanic and Atmospheric Administration, National Ocean Service. [Available online at http://tidesandcurrents.noaa.gov/sltrends/sltrends.shtml]

Southeast and Caribbean

1. Mackun, P., S. Wilson, T. R. Fischetti, and J. Goworowska, 2010: Population Distribution and Change: 2000 to 2010, 12 pp., U.S. Department of Commerce, Economics and Statistics Administration, U.S. Census Bureau. [Available online at http://www.census.gov/prod/cen2010/briefs/c2010br-01.pdf]

2. Ingram, K., K. Dow, and L. Carter, 2012: Southeast Region Technical Report to the National Climate Assessment 334 pp. [Available online at http://downloads.usgcrp.gov/NCA/Activities/NCA_SE_Technical_Report_FINAL_7-23-12.pdf]

3. Portier, C. J., T. K. Thigpen, S. R. Carter, C. H. Dilworth, A. E. Grambsch, J. Gohlke, J. Hess, S. N. Howard, G. Luber, J. T. Lutz, T. Maslak, N. Prudent, M. Radtke, J. P. Rosenthal, T. Rowles, P. A. Sandifer, J. Scheraga, P. J. Schramm, D. Strickman, J. M. Trtanj, and P-Y. Whung, 2010: A Human Health Perspective on Climate Change: A Report Outlining the Research Needs on the Human Health Effects of Climate Change, 80 pp., Environmental Health Perspectives and the National Institute of Environmental Health Services, Research Triangle Park, NC. [Available online at www.niehs.nih.gov/climatereport]

4. Hatfield, J., K. Boote, P. Fay, L. Hahn, C. Izaurralde, B. A. Kimball, T. Mader, J. Morgan, D. Ort, W. Polley, A. Thompson, and D. Wolfe, 2008: Ch. 2: Agriculture. *The Effects of Climate Change on Agriculture, Land Resources, and Biodiversity in the United States. A Report by the U.S. Climate Change Science Program and the Subcommittee on Global Change Research*, P. Backlund, A. Janetos, D. Schimel, J. Hatfield, K. Boote, P. Fay, L. Hahn, C. Izaurralde, B. A. Kimball, T. Mader, J. Morgan, D. Ort, W. Polley, A. Thomson, D. Wolfe, M. G. Ryan, S. R. Archer, R. Birdsey, C. Dahm, L. Heath, J. Hicke, D. Hollinger, T. Huxman, G. Okin, R. Oren, J. Randerson, W. Schlesinger, D. Lettenmaier, D. Major, L. Poff, S. Running, L. Hansen, D. Inouye, B. P. Kelly, L. Meyerson, B. Peterson, and R. Shaw, Eds., U.S. Department of Agriculture, 21-74. [Available online at http://library.globalchange.gov/products/sap-3-4-the-effects-of-climate-change-on-agriculture-land-resources-water-resources-and-biodiversity]

Alexandrov, V. A., and G. Hoogenboom, 2000: Vulnerability and adaptation assessments of agricultural crops under climate change in the Southeastern USA. *Theoretical and Applied Climatology*, 67, 45-63, doi:10.1007/s007040070015.

5. Hallegraeff, G. M., 2010: Ocean climate change, phytoplankton community responses, and harmful algal blooms: A formidable predictive challenge. *Journal of Phycology*, 46, 220-235, doi:10.1111/j.1529-8817.2010.00815.x. [Available online at http://onlinelibrary.wiley.com/doi/10.1111/j.1529-8817.2010.00815.x/full]

Moore, S. K., V. L. Trainer, N. J. Mantua, M. S. Parker, E. A. Laws, L. C. Backer, and L. E. Fleming, 2008: Impacts of climate variability and future climate change on harmful algal blooms and human health. *Environmental Health*, 7, 1-12, doi:10.1186/1476-069X-7-S2-S4. [Available online at http://www.ehjournal.net/content/pdf/1476-069X-7-S2-S4.pdf]

Tester, P. A., R. L. Feldman, A. W. Nau, S. R. Kibler, and R. Wayne Litaker, 2010: Ciguatera fish poisoning and sea surface temperatures in the Caribbean Sea and the West Indies. *Toxicon*, **56**, 698-710, doi:10.1016/j.toxicon.2010.02.026.

Tirado, M. C., R. Clarke, L. A. Jaykus, A. McQuatters-Gollop, and J. M. Frank, 2010: Climate change and food safety: A review. *Food Research International*, **43**, 1745-1765, doi:10.1016/j.foodres.2010.07.003.

Wiedner, C., J. Rücker, R. Brüggemann, and B. Nixdorf, 2007: Climate change affects timing and size of populations of an invasive cyanobacterium in temperate regions. *Oecologia*, **152**, 473-484, doi:10.1007/s00442-007-0683-5.

6. Kunkel, K. E., L. E. Stevens, S. E. Stevens, L. Sun, E. Janssen, D. Wuebbles, C. E. Konrad, II, C. M. Fuhrman, B. D. Keim, M. C. Kruk, A. Billet, H. Needham, M. Schafer, and J. G. Dobson, 2013: Regional Climate Trends and Scenarios for the U.S. National Climate Assessment: Part 2. Climate of the Southeast U.S. NOAA Technical Report 142-2. 103 pp., National Oceanic and Atmospheric Administration, National Environmental Satellite, Data, and Information Service, Washington D.C. [Available online at http://www.nesdis.noaa.gov/technical_reports/NOAA_NESDIS_Tech_Report_142-2-Climate_of_the_Southeast_U.S.pdf]

7. NOAA, cited 2013: Billion Dollar Weather/Climate Disasters, List of Events. National Oceanic and Atmospheric Administration [Available online at http://www.ncdc.noaa.gov/billions/events]

8. Campanella, R., 2010: *Delta Urbanism: New Orleans*. American Planning Association, 224 pp.

9. Strauss, B. H., R. Ziemlinski, J. L. Weiss, and J. T. Overpeck, 2012: Tidally adjusted estimates of topographic vulnerability to sea level rise and flooding for the contiguous United States. *Environmental Research Letters*, **7**, 014033, doi:10.1088/1748-9326/7/1/014033.

10. PRCCC, 2013: State of Puerto Rico's Climate 2010-2013 Executive Summary. Assessing Puerto Rico's Social-Ecological Vulnerabilities in a Changing Climate. ELECTRONIC VERSION, 27 pp., Puerto Rico Climate Change Council. Puerto Rico Coastal Zone Management Program, Department of Natural and Environmental Resources, Office of Ocean and Coastal Resource Management (NOAA-OCRM), San Juan, PR. [Available online at http://www.drna.gobierno.pr/oficinas/arn/recursosvivientes/costasreservasrefugios/pmzc/prccc/prccc-2013/PRCCC_ExecutiveSummary_ElectronicVersion_English.pdf]

11. AWF/AEC/Entergy, 2010: Building a Resilient Energy Gulf Coast: Executive Report, 11 pp., America's Wetland Foundation, America's Energy Coast, and Entergy. [Available online at www.entergy.com/content/our_community/environment/GulfCoastAdaptation/Building_a_Resilient_Gulf_Coast.pdf]

12. State of Louisiana, 2012: Louisiana's Comprehensive Master Plan for a Sustainable Coast, draft Jan 2012, State of Louisiana. Coastal Protection and Restoration Authority, Baton Rouge, LA. [Available online at http://www.coastalmasterplan.louisiana.gov/2012-master-plan/final-master-plan/]

13. DHS, 2011: Louisiana Highway 1/Port Fourchon Study, 76 pp., U.S. Department of Homeland Security. [Available online at http://www.nimsat.org/sites/nimsat/files/Final%20Report.pdf]

14. Obeysekera, J., M. Irizarry, J. Park, J. Barnes, and T. Dessalegne, 2011: Climate change and its implications for water resources management in south Florida. *Stochastic Environmental Research and Risk Assessment*, **25**, 495-516, doi:10.1007/s00477-010-0418-8.

SFWMD: Climate Change and Water Management in South Florida. Interdepartmental Climate Change Group report November 12, 2009. South Florida Water Management District. [Available online at https://my.sfwmd.gov/portal/page/portal/xrepository/sfwmd_repository_pdf/climate_change_and_water_management_in_sflorida_12nov2009.pdf]

15. Berry, L., F. Bloetscher, N. Hernández Hammer, M. Koch-Rose, D. Mitsova-Boneva, J. Restrepo, T. Root, and R. Teegavarapu, 2011: Florida Water Management and Adaptation in the Face of Climate Change, 68 pp., Florida Climate Change Task Force. [Available online at http://floridaclimate.org/docs/water_managment.pdf]

16. UNEP, 2008: Climate Change in the Caribbean and the Challenge of Adaptation, 92 pp., United Nations Environment Programme, Regional Office for Latin America and the Caribbean, Panama City, Panama. [Available online at http://www.pnuma.org/deat1/pdf/Climate_Change_in_the_Caribbean_Final_LOW20oct.pdf]

17. Hammar-Klose, E., and E. Thieler, 2001: National Assessment of Coastal Vulnerability to Future Sea-Level Rise: Preliminary Results for the US Atlantic, Pacific and Gulf of Mexico Coasts. US Reports 99–593, 00-178, and 00-179. U.S. Geological Survey. [Available online at http://woodshole.er.usgs.gov/project-pages/cvi/]

18. Hewes, W., and K. Pitts, 2009: Natural Security: How Sustainable Water Strategies Are Preparing Communities for a Changing Climate, 112 pp., American Rivers, Washington, D.C. [Available online at http://www.americanrivers.org/assets/pdfs/reports-and-publications/natural-security-report.pdf]

19. Devens, T., 2012: Phone Interview. N. Hernández Hammer, recipient

Henderson, B., 2011: Rising Waters Threaten the Coast Of North Carolina. *The Charlotte Observer*, January 18, 2011. The McClatchy Company. [Available online at http://www.charlotteobserver.com/2011/01/18/1983784/rising-waters-threaten-nc-coast.html]

Titus, J., 2002: Does sea level rise matter to transportation along the Atlantic coast? *The Potential Impacts of Climate Change on Transportation, Summary and Discussion Papers, Federal Research Partnership Workshop, October 1-2, 2002*, U.S. Department of Transportation Center for Climate Change and Environmental Forecasting, 135-150.

Midwest

1. Patz, J. A., S. J. Vavrus, C. K. Uejio, and S. L. McLellan, 2008: Climate change and waterborne disease risk in the Great Lakes region of the US. *American Journal of Preventive Medicine*, **35**, 451-458, doi:10.1016/j.amepre.2008.08.026. [Available online at http://www.ajpmonline.org/article/S0749-3797(08)00702-2/fulltext]

2. Winkler, J. A., J. Andresen, J. Bisanz, G. Guentchev, J. Nugent, K. Primsopa, N. Rothwell, C. Zavalloni, J. Clark, H. K. Min, A. Pollyea, and H. Prawiranta, 2013: Ch. 8: Michigan's tart cherry industry: Vulnerability to climate variability and change. *Climate Change in the Midwest: Impacts, Risks, Vulnerability and Adaptation*, S. C. Pryor, Ed., Indiana University Press, 104-116.

3. Vavrus, S., J. E. Walsh, W. L. Chapman, and D. Portis, 2006: The behavior of extreme cold air outbreaks under greenhouse warming. *International Journal of Climatology*, **26**, 1133-1147, doi:10.1002/joc.1301.

4. Gu, L., P. J. Hanson, W. Mac Post, D. P. Kaiser, B. Yang, R. Nemani, S. G. Pallardy, and T. Meyers, 2008: The 2007 eastern US spring freezes: Increased cold damage in a warming world? *BioScience*, **58**, 253-262, doi:10.1641/b580311. [Available online at http://www.jstor.org/stable/10.1641/B580311]

Leakey, A. D. B., 2009: Rising atmospheric carbon dioxide concentration and the future of C$_4$ crops for food and fuel. *Proceedings of the Royal Society B: Biological Sciences*, **276**, 2333-2343, doi:10.1098/rspb.2008.1517. [Available online at http://rspb.royalsocietypublishing.org/content/276/1666/2333.full.pdf+html]

Sage, R., and D. Kubien, 2003: Quo vadis C$_4$? An ecophysiological perspective on global change and the future of C$_4$ plants. *Photosynthesis Research*, **77**, 209-225, doi:10.1023/a:1025882003661.

Lobell, D. B., and C. B. Field, 2007: Global scale climate - crop yield relationships and the impacts of recent warming. *Environmental Research Letters*, **2**, doi:10.1088/1748-9326/2/1/014002. [Available online at http://iopscience.iop.org/1748-9326/2/1/014002/pdf/1748-9326_2_1_014002.pdf]

5. Hatfield, J. L., K. J. Boote, B. A. Kimball, L. H. Ziska, R. C. Izaurralde, D. Ort, A. M. Thomson, and D. Wolfe, 2011: Climate impacts on agriculture: Implications for crop production. *Agronomy Journal*, **103**, 351-370, doi:10.2134/agronj2010.0303.

6. Rosenzweig, C., F. N. Tubiello, R. Goldberg, E. Mills, and J. Bloomfield, 2002: Increased crop damage in the US from excess precipitation under climate change. *Global Environmental Change*, **12**, 197-202, doi:10.1016/S0959-3780(02)00008-0.

7. Pryor, S. C., R. J. Barthelmie, D. T. Young, E. S. Takle, R. W. Arritt, D. Flory, W. J. Gutowski, Jr., A. Nunes, and J. Roads, 2009: Wind speed trends over the contiguous United States. *Journal of Geophysical Research*, **114**, 2169-8996, doi:10.1029/2008JD011416.

8. Ferris, G., cited 2012: State of the Great Lakes 2009. Climate change: Ice duration on the Great Lakes Environment Canada and United States Environmental Protection Agency. [Available online at http://www.epa.gov/solec/sogl2009/]

Mackey, S., 2012: Great Lakes nearshore and coastal systems. *U.S. National Climate Assessment Midwest Technical Input Report*, J. Winkler, J. Andresen, J. Hatfield, D. Bidwell, and D. Brown, Eds., Great Lakes Integrated Sciences and Assessments (GLISA), National Laboratory for Agriculture and the Environment, 14. [Available online at http://glisa.msu.edu/docs/NCA/MTIT_Coastal.pdf]

Wuebbles, D. J., K. Hayhoe, and J. Parzen, 2010: Introduction: Assessing the effects of climate change on Chicago and the Great Lakes. *Journal of Great Lakes Research*, **36**, 1-6, doi:10.1016/j.jglr.2009.09.009.

9. Millerd, F., 2011: The potential impact of climate change on Great Lakes international shipping. *Climatic Change*, **104**, 629-652, doi:10.1007/s10584-010-9872-z.

10. Hellmann, J. J., J. E. Byers, B. G. Bierwagen, and J. S. Dukes, 2008: Five potential consequences of climate change for invasive species. *Conservation Biology*, **22**, 534-543, doi:10.1111/j.1523-1739.2008.00951.x. [Available online at http://onlinelibrary.wiley.com/doi/10.1111/j.1523-1739.2008.00951.x/pdf]

Smith, A. L., N. Hewitt, N. Klenk, D. R. Bazely, N. Yan, S. Wood, I. Henriques, J. I. MacLellan, and C. Lipsig-Mummé, 2012: Effects of climate change on the distribution of invasive alien species in Canada: A knowledge synthesis of range change projections in a warming world. *Environmental Reviews*, **20**, 1-16, doi:10.1139/a11-020.

11. Bai, X., and J. Wang, 2012: Atmospheric teleconnection patterns associated with severe and mild ice cover on the Great Lakes, 1963–2011. *Water Quality Research Journal of Canada* **47**, 421–435, doi:10.2166/wqrjc.2012.009.

12. Maus, E., 2013: Case Studies in Floodplain Regulation, 14 pp. [Available online at http://www.georgetownclimate.org/sites/default/files/Case%20Studies%20in%20Floodplain%20Regulation%206-3-final.pdf]

13. City of Chicago, 2008: City of Chicago Climate Action Plan: Our City. Our Future, 57 pp. [Available online at http://www.chicagoclimateaction.org/filebin/pdf/finalreport/CCAPREPORTFINALv2.pdf]

Great Plains

1. H. John Heinz III Center for Science Economics and the Environment, 2008: *The State of the Nation' Ecosystems 2008: Measuring the Land, Waters, and Living Resources of the United States*, Island Press, 44 pp. [Available online at http://www.heinzctr.org/Ecosystems_files/The%20State%20of%20the%20Nation%27s%20Ecosystems%202008.pdf]

Kostyack, J., J. J. Lawler, D. D. Goble, J. D. Olden, and J. M. Scott, 2011: Beyond reserves and corridors: Policy solutions to facilitate the movement of plants and animals in a changing climate. *Bioscience*, **61**, 713-719, doi:10.1525/bio.2011.61.9.10. [Available online at http://www.bioone.org/doi/pdf/10.1525/bio.2011.61.9.10]

2. Colby, B., and P. Tanimoto, 2011: Using climate information to improve electric utility load forecasting. *Adaptation and Resilience: The Economics of Climate-Water-Energy Challenges in the Arid Southwest*, B. G. Colby, and G. B. Frisvold, Eds., RFF Press, 207-228.

3. Kunkel, K. E., L. E. Stevens, S. E. Stevens, L. Sun, E. Janssen, D. Wuebbles, M. C. Kruk, D. P. Thomas, M. D. Shulski, N. Umphlett, K. G. Hubbard, K. Robbins, L. Romolo, A. Akyuz, T. Pathak, T. R. Bergantino, and J. G. Dobson, 2013: Regional Climate Trends and Scenarios for the U.S. National Climate Assessment: Part 4. Climate of the U.S. Great Plains. NOAA Technical Report NESDIS 142-4. 91 pp., National Oceanic and Atmospheric Administration, National Environmental Satellite, Data, and Information Service, Washington, D.C. [Available online at http://www.nesdis.noaa.gov/technical_reports/NOAA_NESDIS_Tech_Report_142-4-Climate_of_the_U.S.%20Great_Plains.pdf]

4. Trenberth, K. E., J. T. Overpeck, and S. Solomon, 2004: Exploring drought and its implications for the future. *Eos, Transactions, American Geophysical Union*, **85**, 27, doi:10.1029/2004EO030004.

5. Dunnell, K. L., and S. E. Travers, 2011: Shifts in the flowering phenology of the Northern Great Plains: Patterns over 100 years *American Journal of Botany*, **98**, 935-945, doi:10.3732/ajb.1000363. [Available online at http://www.amjbot.org/content/98/6/935.full.pdf+html]

Hu, Q., A. Weiss, S. Feng, and P. S. Baenziger, 2005: Earlier winter wheat heading dates and warmer spring in the U.S. Great Plains. *Agricultural and Forest Meteorology*, **135**, 284-290, doi:10.1016/j.agrformet.2006.01.001.

Wu, C., A. Gonsamo, J. M. Chen, W. A. Kurz, D. T. Price, P. M. Lafleur, R. S. Jassal, D. Dragoni, G. Bohrer, C. M. Gough, S. B. Verma, A. E. Suyker, and J. W. Munger, 2012: Interannual and spatial impacts of phenological transitions, growing season length, and spring and autumn temperatures on carbon sequestration: A North America flux data synthesis. *Global and Planetary Change*, **92-93**, 179-190, doi:10.1016/j.gloplacha.2012.05.021.

6. Nardone, A., B. Ronchi, N. Lacetera, M. S. Ranieri, and U. Bernabucci, 2010: Effects of climate change on animal production and sustainability of livestock systems. *Livestock Science*, **130**, 57-69, doi:10.1016/j.livsci.2010.02.011. [Available online at http://dspace.unitus.it/bitstream/2067/1339/1/LIVSCI%201108%20Nardone%20et%20al%202010.pdf]

Van Dijk, J., N. D. Sargison, F. Kenyon, and P. J. Skuce, 2010: Climate change and infectious disease: Helminthological challenges to farmed ruminants in temperate regions. *Animal*, **4**, 377-392, doi:10.1017/S1751731109990991.

7. NOAA, and USDA, 2008: The Easter Freeze of April 2007: A Climatological Perspective and Assessment of Impacts and Services. NOAA/USDA Tech Report 2008-1, 56 pp., NOAA, U.S. Department of Agriculture. [Available online at http://www1.ncdc.noaa.gov/pub/data/techrpts/tr200801/tech-report-200801.pdf]

8. Konikow, L. F., 2011: Contribution of global groundwater depletion since 1900 to sea-level rise. *Geophysical Research Letters*, **38**, L17401, doi:10.1029/2011GL048604. [Available online at http://onlinelibrary.wiley.com/doi/10.1029/2011GL048604/pdf]

Scanlon, B. R., J. B. Gates, R. C. Reedy, W. A. Jackson, and J. P. Bordovsky, 2010: Effects of irrigated agroecosystems: 2. Quality of soil water and groundwater in the southern High Plains, Texas. *Water Resources Research*, **46**, 1-14, doi:10.1029/2009WR008428. [Available online at http://www.beg.utexas.edu/staffinfo/Scanlon_pdf/Scanlon_et_al_WRR_2010_HP_Irrig_Qual.pdf]

9. Colaizzi, P. D., P. H. Gowda, T. H. Marek, and D. O. Porter, 2009: Irrigation in the Texas High Plains: A brief history and potential reductions in demand. *Journal of Irrigation and Drainage Engineering*, **58**, 257-274, doi:10.1002/ird.418.

10. Oyate Omniciye, 2011: Oglala Lakota Plan, 141 pp. [Available online at http://www.oglalalakotaplan.org/?s=Oglala+Lakota+Plan]

Southwest

1. Theobald, D. M., W. R. Travis, M. A. Drummond, and E. S. Gordon, 2013: Ch. 3: The Changing Southwest. *Assessment of Climate Change in the Southwest United States: A Report Prepared for the National Climate Assessment*, G. Garfin, A. Jardine, R. Merideth, M. Black, and S. LeRoy, Eds., Island Press, 37-55. [Available online at http://swccar.org/sites/all/themes/files/SW-NCA-color-FINALweb.pdf]

2. Baldocchi, D., and S. Wong, 2008: Accumulated winter chill is decreasing in the fruit growing regions of California. *Climatic Change*, **87**, 153-166, doi:10.1007/s10584-007-9367-8.

Battisti, D. S., and R. L. Naylor, 2009: Historical warnings of future food insecurity with unprecedented seasonal heat. *Science*, **323**, 240-244, doi:10.1126/science.1164363.

Lobell, D. B., C. B. Field, K. N. Cahill, and C. Bonfils, 2006: Impacts of future climate change on California perennial crop yields: Model projections with climate and crop uncertainties. *Agricultural and Forest Meteorology*, **141**, 208-218, doi:10.1016/j.agrformet.2006.10.006.

Purkey, D. R., B. Joyce, S. Vicuna, M. W. Hanemann, L. L. Dale, D. Yates, and J. A. Dracup, 2008: Robust analysis of future climate change impacts on water for agriculture and other sectors: A case study in the Sacramento Valley. *Climatic Change*, **87**, 109-122, doi:10.1007/s10584-007-9375-8.

3. Jackson, L., V. R. Haden, S. M. Wheeler, A. D. Hollander, J. Perlman, T. O'Geen, V. K. Mehta, V. Clark, and J. Williams, 2012: Vulnerability and Adaptation to Climate Change in California Agriculture. A White Paper from the California Energy Commission's California Climate Change Center (PIER Program). Publication number: CEC-500-2012-031, 106 pp., Sacramento, California Energy Commission. [Available online at http://www.energy.ca.gov/2012publications/CEC-500-2012-031/CEC-500-2012-031.pdf]

Medellín-Azuara, J., R. E. Howitt, D. J. MacEwan, and J. R. Lund, 2012: Economic impacts of climate-related changes to California agriculture. *Climatic Change*, **109**, 387-405, doi:10.1007/s10584-011-0314-3.

4. Luedeling, E., E. H. Girvetz, M. A. Semenov, and P. H. Brown, 2011: Climate change affects winter chill for temperate fruit and nut trees. PLoS ONE, 6, e20155, doi:10.1371/journal.pone.0020155.

5. Bonfils, C., B. D. Santer, D. W. Pierce, H. G. Hidalgo, G. Bala, T. Das, T. P. Barnett, D. R. Cayan, C. Doutriaux, A. W. Wood, A. Mirin, and T. Nozawa, 2008: Detection and attribution of temperature changes in the mountainous western United States. Journal of Climate, 21, 6404-6424, doi:10.1175/2008JCLI2397.1. [Available online at http://journals.ametsoc.org/doi/abs/10.1175/2008JCLI2397.1]

Williams, A. P., C. D. Allen, C. I. Millar, T. W. Swetnam, J. Michaelsen, C. J. Still, and S. W. Leavitt, 2010: Forest responses to increasing aridity and warmth in the southwestern United States. *Proceedings of the National Academy of Sciences*, **107**, 21289-21294, doi:10.1073/pnas.0914211107. [Available online at http://www.pnas.org/content/107/50/21289.full]

Abatzoglou, J. T., and C. A. Kolden, 2011: Climate change in western US deserts: Potential for increased wildfire and invasive annual grasses. *Rangeland Ecology & Management*, **64**, 471-478, doi:10.2111/rem-d-09-00151.1.

6. Moritz, M. A., M. A. Parisien, E. Batllori, M. A. Krawchuk, J. Van Dorn, D. J. Ganz, and K. Hayhoe, 2012: Climate change and disruptions to global fire activity. *Ecosphere*, **3**, 1-22, doi:10.1890/ES11-00345.1. [Available online at http://www.esajournals.org/doi/pdf/10.1890/ES11-00345.1]

7. Westerling, A. L., H. G. Hidalgo, D. R. Cayan, and T. W. Swetnam, 2006: Warming and earlier spring increase western U.S. forest wildfire activity. *Science*, **313**, 940-943, doi:10.1126/science.1128834.

8. Littell, J. S., D. McKenzie, D. L. Peterson, and A. L. Westerling, 2009: Climate and wildfire area burned in western US ecoprovinces, 1916-2003. *Ecological Applications*, **19**, 1003-1021, doi:10.1890/07-1183.1.

9. Raffa, K. F., B. H. Aukema, B. J. Bentz, A. L. Carroll, J. A. Hicke, M. G. Turner, and W. H. Romme, 2008: Cross-scale drivers of natural disturbances prone to anthropogenic amplification: The dynamics of bark beetle eruptions. *BioScience*, **58,** 501-517, doi:10.1641/b580607. [Available online at http://www.jstor.org/stable/pdfplus/10.1641/B580607.pdf]

10. Gonzalez, P., R. P. Neilson, J. M. Lenihan, and R. J. Drapek, 2010: Global patterns in the vulnerability of ecosystems to vegetation shifts due to climate change. *Global Ecology and Biogeography*, **19,** 755-768, doi:10.1111/j.1466-8238.2010.00558.x. [Available online at http://onlinelibrary.wiley.com/doi/10.1111/j.1466-8238.2010.00558.x/pdf]

    Krawchuk, M. A., M. A. Moritz, M. A. Parisien, J. Van Dorn, and K. Hayhoe, 2009: Global pyrogeography: The current and future distribution of wildfire. *PLoS ONE*, **4,** e5102, doi:10.1371/journal.pone.0005102. [Available online at http://www.plosone.org/article/info%3Adoi/10.1371/journal.pone.0005102]

11. Litschert, S. E., T. C. Brown, and D. M. Theobald, 2012: Historic and future extent of wildfires in the Southern Rockies Ecoregion, USA. *Forest Ecology and Management*, **269,** 124-133, doi:10.1016/j.foreco.2011.12.024.

12. Westerling, A. L., B. P. Bryant, H. K. Preisler, T. P. Holmes, H. G. Hidalgo, T. Das, and S. R. Shrestha, 2011: Climate change and growth scenarios for California wildfire. *Climatic Change*, **109,** 445-463, doi:10.1007/s10584-011-0329-9.

13. Wei, M., H. N. James, B. G. Jeffery, M. Ana, J. Josiah, T. Michael, Y. Christopher, J. Chris, E. M. James, and M. K. Daniel, 2013: Deep carbon reductions in California require electrification and integration across economic sectors. *Environmental Research Letters*, **8,** 014038, doi:10.1088/1748-9326/8/1/014038. [Available online at http://iopscience.iop.org/1748-9326/8/1/014038/pdf/1748-9326_8_1_014038.pdf]

    Wei, M., J. H. Nelson, M. Ting, C. Yang, J. Greenblatt, and J. McMahon, 2012: California's Carbon Challenge. Scenarios for Achieving 80% Emissions Reductions in 2050. Lawrence Berkeley National Laboratory, UC Berkeley, UC Davis, and Itron to the California Energy Commission, UC Berkley, Santa Cruz, CA. [Available online at http://eaei.lbl.gov/sites/all/files/california_carbon_challenge_feb20_20131_0.pdf]

14. Frisvold, G., L. E. Jackson, J. G. Pritchett, and J. Ritten, 2013: Ch. 11: Agriculture and ranching. *Assessment of Climate Change in the Southwest United States: A Report Prepared for the National Climate Assessment*, G. Garfin, A. Jardine, R. Merideth, M. Black, and S. LeRoy, Eds., Island Press, 218-239. [Available online at http://swccar.org/sites/all/themes/files/SW-NCA-color-FINAL.web.pdf]

15. Hoerling, M. P., M. Dettinger, K. Wolter, J. Lukas, J. Eischeid, R. Nemani, B. Liebmann, and K. E. Kunkel, 2013: Ch. 5: Present weather and climate: Evolving conditions *Assessment of Climate Change in the Southwest United States: A Report Prepared for the National Climate Assessment*, G. Garfin, A. Jardine, R. Merideth, M. Black, and S. LeRoy, Eds., Island Press, 74-97. [Available online at http://swccar.org/sites/all/themes/files/SW-NCA-color-FINAL.web.pdf]

Northwest

1. Solecki, W., and C. Rosenzweig, Eds., 2012: U.S. Cities and Climate Change: Urban, Infrastructure, and Vulnerability Issues, Technical Input Report Series, U.S. National Climate Assessment. U.S. Global Change Research Program, Washington, D.C.

2. Lynn, K., O. Grah, P. Hardison, J. Hoffman, E. Knight, A. Rogerson, P. Tillmann, C. Viles, and P. Williams, 2013: Tribal communities. *Climate Change in the Northwest: Implications for Our Landscapes, Waters, And Communities*, P. Mote, M. M. Dalton, and A. K. Snover, Eds., Island Press, 224.

    Voggesser, G., K. Lynn, J. Daigle, F. K. Lake, and D. Ranco, 2013: Cultural impacts to tribes from climate change influences on forests. *Climatic Change*, **120,** 615-626, doi:10.1007/s10584-013-0733-4.

3. Smith, W. B., P. D. Miles, C. H. Perry, and S. A. Pugh, 2009: Forest Resources of the United States, 2007. General Technical Report WO-78. 336 pp., U.S. Department of Agriculture. Forest Service, Washington, D.C. [Available online at http://www.fs.fed.us/nrs/pubs/gtr/gtr_wo78.pdf]

4. Elsner, M. M., L. Cuo, N. Voisin, J. S. Deems, A. F. Hamlet, J. A. Vano, K. E. B. Mickelson, S. Y. Lee, and D. P. Lettenmaier, 2010: Implications of 21st century climate change for the hydrology of Washington State. *Climatic Change*, **102,** 225-260, doi:10.1007/s10584-010-9855-0.

5. Isaak, D. J., S. Wollrab, D. Horan, and G. Chandler, 2011: Climate change effects on stream and river temperatures across the northwest US from 1980–2009 and implications for salmonid fishes. *Climatic Change*, **113,** 499-524, doi:10.1007/s10584-011-0326-z. [Available online at http://link.springer.com/content/pdf/10.1007%2Fs10584-011-0326-z]

6. Hamlet, A. F., M. M. Elsner, G. S. Mauger, S.-Y. Lee, I. Tohver, and R. A. Norheim, 2013: An overview of the Columbia Basin Climate Change Scenarios project: Approach, methods, and summary of key results. *Atmosphere-Ocean*, **51,** 392-415, doi:10.1080/07055900.2013.819555. [Available online at http://www.tandfonline.com/doi/pdf/10.1080/07055900.2013.819555]

7. Eidenshink, J., B. Schwind, K. Brewer, Z. Zhu, B. Quayle, and S. Howard, 2007: A project for monitoring trends in burn severity. *Fire Ecology*, **3,** 3-21. [Available online at http://fireecology.org/docs/Journal/pdf/Volume03/Issue01/003.pdf]

    USGS, cited 2012: National Monitoring Trends in Burn Severity (MTBS) Burned Area Boundaries Dataset. U.S. Geological Survey. [Available online at http://www.mtbs.gov/compositfire/mosaic/bin-release/burnedarea.html]

8. USFS, cited 2012: Forest Service, Insect & Disease Detection Survey Data Explorer. U.S. Department of Agriculture, U.S. Forest Service. [Available online at http://foresthealth.fs.usda.gov/portal]

9. NRC, 2011: Ch. 5: Impacts in the next few decades and coming centuries. *Climate Stabilization Targets: Emissions, Concentrations, and Impacts over Decades to Millennia*, Committee on Stabilization Targets for Atmospheric Greenhouse Gas Concentration, The National Academies Press, 298. [Available online at http://www.nap.edu/catalog.php?record_id=12877]

10. Bailey, R. G., 1995: Description of the Ecoregions of the United States (2nd ed.). 1995. Misc Pub. No. 1391. U.S. Department of Agriculture, Forest Service. [Available online at http://nationalatlas.gov/mld/ecoregp.html]

11. Littell, J. S., E. E. Oneil, D. McKenzie, J. A. Hicke, J. A. Lutz, R. A. Norheim, and M. M. Elsner, 2010: Forest ecosystems, disturbance, and climatic change in Washington State, USA. *Climatic Change*, **102,** 129-158, doi:10.1007/s10584-010-9858-x.

REFERENCES

12. USFWS, 2010: Rising to the Urgent Challenge: Strategic Plan for Responding to Accelerating Climate Change, 32 pp., U.S. Fish and Wildlife Service, U.S. Department of the Interior, Washington, D.C. [Available online at http://www.fws.gov/home/climatechange/pdf/CCStrategicPlan.pdf]

Alaska

1. Stewart, B. C., K. E. Kunkel, L. E. Stevens, L. Sun, and J. E. Walsh, 2013: Regional Climate Trends and Scenarios for the U.S. National Climate Assessment: Part 7. Climate of Alaska. NOAA Technical Report NESDIS 142-7. 60 pp. [Available online at http://www.nesdis.noaa.gov/technical_reports/NOAA_NESDIS_Tech_Report_142-7-Climate_of_Alaska.pdf]

2. Bieniek, P. A., J. E. Walsh, R. L. Thoman, and U. S. Bhatt, 2014: Using climate divisions to analyze variations and trends in Alaska temperature and precipitation. *Journal of Climate*, **in press**, doi:10.1175/JCLI-D-13-00342.1. [Available online at http://journals.ametsoc.org/doi/pdf/10.1175/JCLI-D-13-00342.1]

   Wendler, G., L. Chen, and B. Moore, 2012: The first decade of the new century: A cooling trend for most of Alaska. *The Open Atmospheric Science Journal*, **6**, 111-116, doi:10.2174/1874282301206010111. [Available online at http://benthamscience.com/open/toascj/articles/V006/111TOASCJ.pdf]

3. CCSP, 2008: Weather and Climate Extremes in a Changing Climate - Regions of Focus - North America, Hawaii, Caribbean, and U.S. Pacific Islands. A Report by the U.S. Climate Change Science Program and the Subcommittee on Global Change Research. T. R. Karl, G. A. Meehl, C. D. Miller, S. J. Hassol, A. M. Waple, and W. L. Murray, Eds., 164 pp., Department of Commerce, NOAA's National Climatic Data Center, Washington, D.C. [Available online at http://downloads.globalchange.gov/sap/sap3-3/sap3-3-final-all.pdf]

4. BIA, cited 2012: Alaska Region Overview. U.S. Department of the Interior, Bureau of Indian Affairs. [Available online at http://www.bia.gov/WhoWeAre/RegionalOffices/Alaska/]

5. Maslowski, W., J. Clement Kinney, M. Higgins, and A. Roberts, 2012: The future of Arctic sea ice. *Annual Review of Earth and Planetary Sciences*, **40**, 625-654, doi:10.1146/annurev-earth-042711-105345. [Available online at http://www.annualreviews.org/doi/pdf/10.1146/annurev-earth-042711-105345]

6. Stroeve, J. C., M. C. Serreze, M. M. Holland, J. E. Kay, J. Malanik, and A. P. Barrett, 2012: The Arctic's rapidly shrinking sea ice cover: A research synthesis. *Climatic Change*, **110**, 1005-1027, doi:10.1007/s10584-011-0101-1. [Available online at http://link.springer.com/content/pdf/10.1007%2Fs10584-011-0101-1.pdf]

7. Stroeve, J., M. M. Holland, W. Meier, T. Scambos, and M. Serreze, 2007: Arctic sea ice decline: Faster than forecast. *Geophysical Research Letters*, **34**, L09501, doi:10.1029/2007GL029703. [Available online at http://www.agu.org/pubs/crossref/2007/2007GL029703.shtml]

   Wang, M., and J. E. Overland, 2009: A sea ice free summer Arctic within 30 years? *Geophysical Research Letters*, **36**, L07502, doi:10.1029/2009GL037820. [Available online at http://onlinelibrary.wiley.com/doi/10.1029/2009GL037820/pdf]

   ———, 2012: A sea ice free summer Arctic within 30 years: An update from CMIP5 models. *Geophysical Research Letters*, **39**, L18501, doi:10.1029/2012GL052868. [Available online at http://onlinelibrary.wiley.com/doi/10.1029/2012GL052868/pdf]

8. Jorgenson, T., K. Yoshikawa, M. Kanevskiy, Y. Shur, V. Romanovsky, S. Marchenko, G. Grosse, J. Brown, and B. Jones, 2008: Permafrost characteristics of Alaska. *Extended Abstracts of the Ninth International Conference on Permafrost, June 29-July 3, 2008.* , D. L. Kane, and K. M. Hinkel, Eds., University of Alaska Fairbanks, 121-123. [Available online at http://permafrost.gi.alaska.edu/sites/default/files/AlaskaPermafrostMap_Front_Dec2008_Jorgenson_etal_2008.pdf]

9. Romanovsky, V. E., S. S. Marchenko, R. Daanen, D. O. Sergeev, and D. A. Walker, 2008: Soil climate and frost heave along the Permafrost/Ecological North American Arctic Transect. *Proceedings of the Ninth International Conference on Permafrost*, Institute of Northern Engineering, University of Alaska Fairbanks, 1519-1524 pp.

10. French, H., 2011: Geomorphic change in northern Canada. *Changing Cold Environments: A Canadian Perspective*, H. French, and O. Slaymaker, Eds., John Wiley & Sons, Ltd, 200-221.

    Romanovsky, V. E., S. L. Smith, and H. H. Christiansen, 2010: Permafrost thermal state in the polar Northern Hemisphere during the international polar year 2007-2009: A synthesis. *Permafrost and Periglacial Processes*, **21**, 106-116, doi:10.1002/ppp.689. [Available online at http://onlinelibrary.wiley.com/doi/10.1002/ppp.689/pdf]

11. Avis, C. A., A. J. Weaver, and K. J. Meissner, 2011: Reduction in areal extent of high-latitude wetlands in response to permafrost thaw. *Nature Geoscience*, **4**, 444-448, doi:10.1038/ngeo1160.

    Euskirchen, E. S., A. D. McGuire, D. W. Kicklighter, Q. Zhuang, J. S. Clein, R. J. Dargaville, D. G. Dye, J. S. Kimball, K. C. McDonald, J. M. Melillo, V. E. Romanovsky, and N. V. Smith, 2006: Importance of recent shifts in soil thermal dynamics on growing season length, productivity, and carbon sequestration in terrestrial high-latitude ecosystems. *Global Change Biology*, **12**, 731-750, doi:10.1111/j.1365-2486.2006.01113.x. [Available online at http://onlinelibrary.wiley.com/doi/10.1111/j.1365-2486.2006.01113.x/pdf]

    Lawrence, D. M., and A. G. Slater, 2008: Incorporating organic soil into a global climate model. *Climate Dynamics*, **30**, 145-160, doi:10.1007/s00382-007-0278-1. [Available online at http://www.springerlink.com/index/10.1007/s00382-007-0278-1]

12. Jafarov, E. E., S. S. Marchenko, and V. E. Romanovsky, 2012: Numerical modeling of permafrost dynamics in Alaska using a high spatial resolution dataset. *The Cryosphere Discussions*, **6**, 89-124, doi:10.5194/tcd-6-89-2012.

13. ICLEI, cited 2013: Homer, Alaska's Climate Adaptation Progress Despite Uncertainties. ICLEI. [Available online at http://www.cakex.org/virtual-library/2555]

14. Bronen, R., 2011: Climate-induced community relocations: Creating an adaptive governance framework based in human rights doctrine. *NYU Review Law & Social Change*, **35**, 357-408. [Available online at http://socialchangenyu.files.wordpress.com/2012/08/climate-induced-migration-bronen-35-2.pdf]

Hawai'i and Pacific Islands

1. Döll, P., 2002: Impact of climate change and variability on irrigation requirements: A global perspective. *Climatic Change*, **54**, 269-293, doi:10.1023/A:1016124032231.

   Sivakumar, M. V. K., and J. Hansen, 2007: *Climate Prediction and Agriculture: Advances and Challenges.* Springer, 307 pp.

Wairiu, M., M. Lal, and V. Iese, 2012: Ch. 5: Climate change implications for crop production in Pacific Islands region. *Food Production - Approaches, Challenges and Tasks*, A. Aladjadjiyan, Ed. [Available online at http://www.intechopen.com/books/food-production-approaches-challenges-and-tasks/climate-change-implications-for-crop-production-in-pacific-islands-region]

2. Barnett, J., and W. N. Adger, 2003: Climate dangers and atoll countries. *Climatic Change*, **61**, 321-337, doi:10.1023/B:CLIM.0000004559.08755.88.

3. Easterling, W. E., P. K. Aggarwal, P. Batima, K. M. Brander, L. Erda, S. M. Howden, A. Kirilenko, J. Morton, J.-F. Soussana, J. Schmidhuber, and F. N. Tubiello, 2007: Ch. 5: Food, fibre, and forest products. *Climate Change 2007: Impacts, Adaptation and Vulnerability. Contribution of Working Group II to the Fourth Assessment Report of the Intergovernmental Panel on Climate Change*, M. L. Parry, O. F. Canziani, J. P. Palutikof, P. J. Van der Linden, and C. E. Hanson, Eds., Cambridge University Press, 273-313. [Available online at http://www.ipcc.ch/pdf/assessment-report/ar4/wg2/ar4-wg2-chapter5.pdf]

4. Maclellan, N., 2009: Rising tides—responding to climate change in the Pacific. *Social Alternatives*, **28**, 8-13.

5. Benning, T. L., D. LaPointe, C. T. Atkinson, and P. M. Vitousek, 2002: Interactions of climate change with biological invasions and land use in the Hawaiian Islands: Modeling the fate of endemic birds using a geographic information system, *Proceedings of the National Academy of Sciences*, **99**, 14246-14249, doi:10.1073/pnas.162372399. [Available online at http://www.pnas.org/content/99/22/14246.full.pdf+html]

LaPointe, D. A., C. T. Atkinson, and M. D. Samuel, 2012: Ecology and conservation biology of avian malaria. *Annals of the New York Academy of Sciences*, **1249**, 211-226, doi:10.1111/j.1749-6632.2011.06431.x.

6. Gilman, E. L., J. Ellison, N. C. Duke, and C. Field, 2008: Threats to mangroves from climate change and adaptation options: A review. *Aquatic Botany*, **89**, 237-250, doi:10.1016/j.aquabot.2007.12.009.

7. Waycott, M., L. McKenzie, J. E. Mellors, J. C. Ellison, M. T. Sheaves, C. Collier, A. M. Schwarz, A. Webb, J. E. Johnson, and C. E. Payri, 2011: Ch. 6: Vulnerability of mangroves, seagrasses and intertidal flats in the tropical Pacific to climate change. *Vulnerability of Tropical Pacific Fisheries and Aquaculture to Climate Change*, J. D. Bell, J. E. Johnson, and A. J. Hobday, Eds., Secretariat of the Pacific Community, 297-368.

8. Arata, J. A., P. R. Sievert, and M. B. Naughton, 2009: Status assessment of Laysan and black-footed albatrosses, North Pacific Ocean, 1923-2005. Scientific Investigations Report 2009-5131, 80 pp., U.S. Geological Survey. [Available online at http://pubs.usgs.gov/sir/2009/5131/pdf/sir20095131.pdf]

9. Waikīkī Improvement Association, 2008: Economic Impact Analysis of the Potential Erosion of Waikīkī Beach, 123 pp., Hospitality Advisors, LLC, Honolulu, HI. [Available online at http://the.honoluluadvertiser.com/dailypix/2008/Dec/07/HospitalityAdvisorsReport.pdf]

10. Lewis, N., 2012: Islands in a sea of change: Climate change, health and human security in small island states. *National Security and Human Health Implications of Climate Change*, H. J. S. Fernando, Z. Klaić, and J. L. McCulley, Eds., Springer, 13-24.

11. Merrifield, M. A., 2011: A shift in western tropical Pacific sea level trends during the 1990s. *Journal of Climate*, **24**, 4126-4138, doi:10.1175/2011JCLI3932.1. [Available online at http://journals.ametsoc.org/doi/pdf/10.1175/2011JCLI3932.1]

12. Pratchett, M. S., P. L. Munday, N. A. J. Graham, M. Kronen, S. Pinca, K. Friedman, T. D. Brewer, J. D. Bell, S. K. Wilson, J. E. Cinner, J. P. Kinch, R. J. Lawton, A. J. Williams, L. Chapman, F. Magron, and A. Webb, 2011: Ch. 9: Vulnerability of coastal fisheries in the tropical Pacific to climate change: Summary for Pacific Island Countries and Territories. *Vulnerability of Tropical Pacific Fisheries and Aquaculture to Climate Change*, J. D. Bell, J. E. Johnson, and A. J. Hobday, Eds., Secretariat of the Pacific Community, 367-370.

13. NOAA, 2010: Adapting to Climate Change: A Planning Guide for State Coastal Managers, 133 pp., NOAA Office of Ocean and Coastal Resource Management, Silver Spring, MD. [Available online at http://coastalmanagement.noaa.gov/climate/docs/adaptationguide.pdf]

14. HDLNR, 2011: The Rain Follows The Forest: A Plan to Replenish Hawaii's Source of Water, 24 pp., Department of Land and Natural Resources, State of Hawai`i. [Available online at http://dlnr.hawaii.gov/rain/files/2014/02/The-Rain-Follows-the-Forest.pdf]

Rural Communities

1. HRSA, cited 2012: Defining the Rural Population. U.S. Department of Health and Human Services, Health Resources and Services Administration. [Available online at http://www.hrsa.gov/ruralhealth/policy/definition_of_rural.html]

U.S. Census Bureau, cited 2012: United States Census 2010. U.S. Census Bureau,. [Available online at http://www.census.gov/2010census/]

——, cited 2012: 2010 Census Urban and Rural Classification and Urban Area Criteria. U.S. Census Bureau, U.S. Department of Commerce. [Available online at http://www.census.gov/geo/reference/frn.html]

USDA, cited 2012: Atlas of Rural and Small-Town America. U.S. Department of Agriculture, Economic Research Service. [Available online at http://www.ers.usda.gov/data-products/atlas-of-rural-and-small-town-america/go-to-the-atlas.aspx]

2. ERS, cited 2012: Economic Research Service, U.S. Department of Agriculture. Economic Research Service, U.S. Department of Agriculture. [Available online at http://www.ers.usda.gov/briefing/rurality/newdefinitions/]

3. Peterson, T. C., P. A. Stott, and S. Herring, 2012: Explaining extreme events of 2011 from a climate perspective. *Bulletin of the American Meteorological Society*, **93**, 1041-1067, doi:10.1175/BAMS-D-12-00021.1. [Available online at http://journals.ametsoc.org/doi/pdf/10.1175/BAMS-D-12-00021.1]

4. DOT, cited 2010: Freight Analysis Framework (Version 3) Data Tabulation Tool, Total Flows. U.S. Department of Transportation. [Available online at http://faf.ornl.gov/fafweb/Extraction1.aspx]

5. Kunkel, K. E., D. R. Easterling, K. Hubbard, and K. Redmond, 2009: 2009 update to data originally published in "Temporal variations in frost-free season in the United States: 1895–2000". *Geophysical Research Letters*, **31**, L03201, doi:10.1029/2003GL018624. [Available online at http://onlinelibrary.wiley.com/doi/10.1029/2003GL018624/full]

6. Westerling, A. L., H. G. Hidalgo, D. R. Cayan, and T. W. Swetnam, 2006: Warming and earlier spring increase western U.S. forest wildfire activity. *Science*, **313**, 940-943, doi:10.1126/science.1128834.

7. USDA, cited 2013: Atlas of Rural and Small-Town America. U.S. Department of Agriculture, Economic Research Service. [Available online at http://www.ers.usda.gov/data-products/atlas-of-rural-and-small-town-america/go-to-the-atlas.aspx]

8. Allen, C. D., C. Birkeland, I. Chapin. F.S., P. M. Groffman, G. R. Guntenspergen, A. K. Knapp, A. D. McGuire, P. J. Mulholland, D. P. C. Peters, D. D. Roby, and G. Sugihara, 2009: Thresholds of Climate Change in Ecosystems: Final Report, Synthesis and Assessment Product 4.2, 172 pp., U.S. Geological Survey, University of Nebraska Lincoln. [Available online at http://digitalcommons.unl.edu/cgi/viewcontent.cgi?article=1009&context=usgspubs]

Staudinger, M. D., N. B. Grimm, A. Staudt, S. L. Carter, F. S. Chapin, III, P. Kareiva, M. Ruckelshaus, and B. A. Stein, 2012: Impacts of Climate Change on Biodiversity, Ecosystems, and Ecosystem Services. Technical Input to the 2013 National Climate Assessment 296 pp., U.S. Geological Survey, Reston, VA. [Available online at http://downloads.usgcrp.gov/NCA/Activities/Biodiversity-Ecosystems-and-Ecosystem-Services-Technical-Input.pdf]

9. Pietrowsky, R., D. Raff, C. McNutt, M. Brewer, T. Johnson, T. Brown, M. Ampleman, C. Baranowski, J. Barsugli, L. D. Brekke, L. Brekki, M. Crowell, D. Easterling, A. Georgakakos, N. Gollehon, J. Goodrich, K. A. Grantz, E. Greene, P. Groisman, R. Heim, C. Luce, S. McKinney, R. Najjar, M. Nearing, D. Nover, R. Olsen, C. Peters-Lidard, L. Poff, K. Rice, B. Rippey, M. Rodgers, A. Rypinski, M. Sale, M. Squires, R. Stahl, E. Z. Stakhiv, and M. Strobel, 2012: Water Resources Sector Technical Input Report in Support of the U.S. Global Change Research Program, National Climate Assessment - 2013, 31 pp.

10. Lal, P., J. R. R. Alavalapati, and E. D. Mercer, 2011: Socio-economic impacts of climate change on rural United States. *Mitigation and Adaptation Strategies for Global Change*, **16**, 819-844, doi:10.1007/s11027-011-9295-9. [Available online at http://www.srs.fs.usda.gov/pubs/ja/2011/ja_2011_lal_002.pdf]

11. Burkett, V., and M. Davidson, 2012: *Coastal Impacts, Adaptation and Vulnerabilities: A Technical Input to the 2013 National Climate Assessment.* Island Press, 216 pp.

Hoyos, C. D., P. A. Agudelo, P. J. Webster, and J. A. Curry, 2006: Deconvolution of the factors contributing to the increase in global hurricane intensity. *Science*, **312**, 94-97, doi:10.1126/science.1123560. [Available online at http://www.jstor.org/stable/3845986?origin=JSTOR-pdf]

Rygel, L., D. O'Sullivan, and B. Yarnal, 2006: A method for constructing a Social Vulnerability Index: An application to hurricane storm surges in a developed country. *Mitigation and Adaptation Strategies for Global Change*, **11**, 741-764, doi:10.1007/s11027-006-0265-6. [Available online at http://www.cara.psu.edu/about/publications/Rygel_et_al_MASGC.pdf]]

Wu, S. Y., B. Yarnal, and A. Fisher, 2002: Vulnerability of coastal communities to sea-level rise: A case study of Cape May County, New Jersey, USA. *Climate Research*, **22**, 255-270, doi:10.3354/cr022255.

12. Galgano, F. A., and B. C. Douglas, 2000: Shoreline position prediction: Methods and errors. *Environmental Geosciences*, **7**, 23-31, doi:10.1046/j.1526-0984.2000.71006.x.

13. Isserman, A. M., E. Feser, and D. E. Warren, 2009: Why some rural places prosper and others do not. *International Regional Science Review*, **32**, 300-342, doi:10.1177/0160017609336090.

14. Kraybill, D. S., and L. Lobao, 2001: The Emerging Roles of County Governments in Rural America: Findings from a Recent National Survey. American Agricultural Economics Association (New Name 2008: Agricultural and Applied Economics Association), 20 pp. [Available online at http://ageconsearch.umn.edu/bitstream/20697/1/sp01kr01.pdf]

15. Berkes, F., 2007: Understanding uncertainty and reducing vulnerability: Lessons from resilience thinking. *Natural Hazards*, **41**, 283-295, doi:10.007/s11069-006-9036-7.

Nelson, D. R., 2011: Adaptation and resilience: Responding to a changing climate. *Wiley Interdisciplinary Reviews: Climate Change*, **2**, 113-120, doi:10.1002/wcc.91. [Available online at http://onlinelibrary.wiley.com/doi/10.1002/wcc.91/pdf]

Ostrom, E., 2009: A general framework for analyzing sustainability of social-ecological systems. *Science*, **325**, 419-422, doi:10.1126/science.1172133. [Available online at http://www.era-mx.org/biblio/Ostrom,%202009.pdf]

Coasts

1. Houston, J. R., 2008: The economic value of beaches – a 2008 update. *Shore & Beach*, **76**, 22-26.

2. OTTI, 2012: Overseas Visitation Estimates for U.S. States, Cities, and Census Regions: 2011, 6 pp., U.S. Department of Commerce, International Trade Commission, Office of Travel and Tourism Industries, Washington, D.C. [Available online at http://tinet.ita.doc.gov/outreachpages/download_data_table/2011_States_and_Cities.pdf]

3. Moser, S. C., S. J. Williams, and D. F. Boesch, 2012: Wicked challenges at land's end: Managing coastal vulnerability under climate change. *Annual Review of Environment and Natural Resources*, **37**, 51-78, doi:10.1146/annurev-environ-021611-135158. [Available online at http://susannemoser.com/documents/Moseretal_2012_AnnualReview_preformat.pdf]

4. GAO, 2006: Natural Gas: Factors Affecting Prices and Potential Impacts on Consumers (GAO-06-420T). 28 pp., U.S. Government Accountability Office, Washington, D.C. [Available online at http://www.gao.gov/assets/120/112796.pdf]

5. FHWA, 2008: Highways in the Coastal Environment, Second Edition. Hydraulic Engineering Circular No. 25. FHWA-NHI-07-096. S. L. K. Douglass, J., Ed., 250 pp., Federal Highway Administration. Department of Civil Engineering, University of South Alabama, Mobile, AL. [Available online at http://www.fhwa.dot.gov/engineering/hydraulics/pubs/07096/07096.pdf]

FDEP, 2012: Critically Eroded Beaches in Florida, 76 pp., Florida Department of Envrionmental Protection, Bureau of Beaches and Coastal Systems, Division of Water Resource Management. [Available online at http://www.dep.state.fl.us/beaches/publications/pdf/critical-erosion-report-2012.pdf]

Texas General Land Office, cited 2012: Caring for the Coast: Coastal Management Program. State of Texas. [Available online at http://www.glo.texas.gov/what-we-do/caring-for-the-coast/grants-funding/cmp/index.html]

Wolshon, B., 2006: Evacuation planning and engineering for Hurricane Katrina. *The Bridge*, **36,** 27-34.

6. California King Tides Initiative, cited 2012: California King Tides Initiative. [Available online at http://www.californiakingtides.org/aboutus/]

State of Washington, cited 2012: Climate Change, King Tides in Washington State. Department of Ecology, State of Washington. [Available online at http://www.ecy.wa.gov/climatechange/ipa_hightide.htm]

Turner, S., 2011: Extreme high tides expected along RI coast; grab your camera *East Greenwich Patch*, October 25, 2011. [Available online at http://eastgreenwich.patch.com/articles/extreme-high-tides-expected-along-ri-coast-b5b7ee05]

Watson, S., 2011: Alignment of the Sun, moon and Earth will cause unusually high tides. *pressofAtlanticCity.com*, October 25, 2011. [Available online at http://www.pressofatlanticcity.com/news/top_three/sun-moon-earth-line-up-for-unusually-high-tides/article_4080f60a-ff70-11e0-ab3e-001cc4c002e0.html]

7. MDOT, cited 2003: Bridge Design Guide. Maine Department of Transportation, Prepared by Guertin Elhorton & Associates. [Available online at http://www.maine.gov/mdot/technicalpubs/bdg.htm]

8. Bierwagen, B. G., D. M. Theobald, C. R, Pyke, A Choate, P. Groth, J V. Thomas, and P. Morefield, 2010: National housing and impervious surface scenarios for integrated climate impact assessments. *Proceedings of the National Academy of Sciences*, **107,** 20887-20892, doi:10.1073/pnas.1002096107.

Bjerklie, D. M., J. R. Mullaney, J. R. Stone, B. J. Skinner, and M. A. Ramlow, 2012: Preliminary Investigation of the Effects of Sea-Level Rise on Groundwater Levels in New Haven, Connecticut. U.S. Geological Survey Open-File Report 2012-1025, 56 pp., U.S. Department off the Interior and U.S. Geological Survey. [Available online at http://pubs.usgs.gov/of/2012/1025/pdf/ofr2012-1025_report_508.pdf]

Changnon, S. A., 2011: Temporal distribution of weather catastrophes in the USA. *Climatic Change*, **106,** 129-140, doi:10.1007/s10584-010-9927-1.

Johnson, L., 2012: Rising groundwater may flood underground infrastructure of coastal cities *Scientific American*, May 2, 2012. Nature Publishing Group. [Available online at http://www.scientificamerican.com/article.cfm?id=rising-groundwater-may-flood-underground-infrastructure-of-coastal-cities]

Peterson, T. C., P. A. Stott, and S. Herring, 2012: Explaining extreme events of 2011 from a climate perspective. *Bulletin of the American Meteorological Society*, **93,** 1041-1067, doi:10.1175/BAMS-D-12-00021.1. [Available online at http://journals.ametsoc.org/doi/pdf/10.1175/BAMS-D-12-00021.1]

Seneviratne, S. I., N. Nicholls, D. Easterling, C. M. Goodess, S. Kanae, J. Kossin, Y. Luo, J. Marengo, K. McInnes, M. Rahimi, M. Reichstein, A. Sorteberg, C. Vera, and X. Zhang, 2012: Ch. 3: Changes in climate extremes and their impacts on the natural physical environment. *Managing the Risks of Extreme Events and Disasters to Advance Climate Change Adaptation. A Special Report of Working Groups I and II of the Intergovernmental Panel on Climate Change (IPCC)*, C. B. Field, V. Barros, T. F. Stocker, Q. Dahe, D. J. Dokken, K. L. Ebi, M. D. Mastrandrea, K. J. Mach, G.-K. Plattner, S. K.

Allen, M. Tignor, and P. M. Midgley, Eds., Cambridge University Press, 109-230.

Toll, D. G., cited 2012: The Impact of Changes in the Water Table and Soil Moisture on Structural Stability of Buildings and Foundation Systems. Systematic review CEE10-005 (SR90). Collaboration for Environmental Evidence. [Available online at http://www.environmentalevidence.org/Documents/Draft_reviews/Draftreview10-005.pdf]

9. Rotzoll, K., and C. H. Fletcher, 2013: Assessment of groundwater inundation as a consequence of sea-level rise. *Nature Climate Change*, **3,** 477-481, doi:10.1038/nclimate1725.

Solecki, W., and C. Rosenzweig, Eds., 2012: U.S. Cities and Climate Change: Urban, Infrastructure, and Vulnerability Issues, Technical Input Report Series, U.S. National Climate Assessment. U.S. Global Change Research Program, Washington, D.C.

Hilton, T. W., R. G. Najjar, L. Zhong, and M. Li, 2008: Is there a signal of sea-level rise in Chesapeake Bay salinity? *Journal of Geophysical Research: Oceans*, **113,** C09002, doi:10.1029/2007jc004247. [Available online at http://onlinelibrary.wiley.com/doi/10.1029/2007JC004247/pdf]

10. CCAP and EESI, 2012: Climate Adaptation & Transportation: Identifying Information and Assistance Needs, 66 pp., Center for Clean Air Policy and Environmental and Energy Study Institute, Washington, D.C. [Available online at http://cakex.org/virtual-library/climate-adaptation-transportation-identifying-information-and-assistance-needs]

EPA, 2008: A Screening Assessment of the Potential Impacts of Climate Change on Combined Sewer Overflow (CSO) Mitigation in the Great Lakes and New England Regions. EPA/600/R-07/033F, 50 pp., U.S. Environmental Protection Agency, Washington, D.C. [Available online at http://ofmpub.epa.gov/eims/eimscomm.getfile?p_download_id=472009]

Kenward, A., D. Yawitz, and U. Raja, 2013: Sewage Overflows From Hurricane Sandy, 43 pp., Climate Central. [Available online at http://www.climatecentral.org/pdfs/Sewage.pdf]

11. Hayhoe, K., M. Robson, J. Rogula, M. Auffhammer, N. Miller, J. VanDorn, and D. Wuebbles, 2010: An integrated framework for quantifying and valuing climate change impacts on urban energy and infrastructure: A Chicago case study. *Journal of Great Lakes Research*, **36,** 94-105, doi:10.1016/j.jglr.2010.03.011.

Perez, P. R., 2009: *Potential Impacts of Climate Change on California's Energy Infrastructure and Identification of Adaptation Measures: Staff Paper.* California Energy Commission, 23 pp.

Sathaye, J., L. Dale, P. Larsen, G. Fitts, K. Koy, S. Lewis, and A. Lucena, 2011: Estimating Risk to California Energy Infrastructure from Projected Climate Change, 85 pp., Ernest Orlando Lawrence Berkeley National Laboratory, California Energy Commission, Berkeley, CA. [Available online at http://www.osti.gov/bridge/servlets/purl/1026011/1026011.PDF]

Wilbanks, T., S. Fernandez, G. Backus, P. Garcia, K. Jonietz, P. Kirshen, M. Savonis, B. Solecki, and L. Toole, 2012: Climate Change and Infrastructure, Urban Systems, and Vulnerabilities. Technical Report to the U.S. Department of Energy in Support of the National Climate Assessment, 119 pp., Oak Ridge National Laboratory. U.S Department of Energy, Office of Science, Oak

REFERENCES

Ridge, TN. [Available online at http://www.esd.ornl.gov/eess/Infrastructure.pdf]

12. DOT, 2012: Climate Impacts and U.S. Transportation: Technical Input Report for the National Climate Assessment. DOT OST/P-33.

13. Francis, R. A., S. M. Falconi, R. Nateghi, and S. D. Guikema, 2011: Probabilistic life cycle analysis model for evaluating electric power infrastructure risk mitigation investments. *Climatic Change*, **106**, 31-55, doi:10.1007/s10584-010-0001-9.

    Rosato, V., L. Issacharoff, F. Tiriticco, S. Meloni, S. Porcellinis, and R. Setola, 2008: Modelling interdependent infrastructures using interacting dynamical models. *International Journal of Critical Infrastructures*, **4**, 63-79, doi:10.1504/IJCIS.2008.016092.

    Vugrin, E. D., and R. C. Camphouse, 2011: Infrastructure resilience assessment through control design. *International Journal of Critical Infrastructures*, **7**, 243-260, doi:10.1504/IJCIS.2011.042994.

    Zimmerman, R., 2006: Ch. 34: Critical infrastructure and interdependency. *The McGraw-Hill Homeland Security Handbook*, D. G. Kamien, Ed., McGraw-Hill, 523-545.

    Vugrin, E. D., D. E. Warren, M. A. Ehlen, and R. C. Camphouse, 2010: A framework for assessing the resilience of infrastructure and economic systems. *Sustainable and Resilient Critical Infrastructure Systems*, K. Gopalakrishnan, and S. Peeta, Eds., Springer Berlin Heidelberg, 77-116.

14. Hallegatte, S., 2008: Adaptation to climate change: Do not count on climate scientists to do your work, 15 pp. [Available online at http://regulation2point0.org/wp-content/uploads/downloads/2010/04/RP08-01_topost.pdf]

    U.S. Government, 2009: Executive Order 13514. Federal Leadership in Environmental, Energy, and Economic Performance. 52117-52127 pp., Federal Register, Washington, D.C. [Available online at http://www.whitehouse.gov/assets/documents/2009fedleader_eo_rel.pdf]

15. Parris, A., P. Bromirski, V. Burkett, D. Cayan, M. Culver, J. Hall, R. Horton, K. Knuuti, R. Moss, J. Obeysekera, A. Sallenger, and J. Weiss, 2012: Global Sea Level Rise Scenarios for the United States National Climate Assessment. NOAA Tech Memo OAR CPO-1, 37 pp., National Oceanic and Atmospheric Administration, Silver Spring, MD. [Available online at http://scenarios.globalchange.gov/sites/default/files/NOAA_SLR_r3_0.pdf]

16. Neumann, J., D. Hudgens, J. Herter, and J. Martinich, 2010: The economics of adaptation along developed coastlines. *Wiley Interdisciplinary Reviews: Climate Change*, **2**, 89-98, doi:10.1002/wcc.90. [Available online at http://onlinelibrary.wiley.com/doi/10.1002/wcc.90/pdf]

17. Neumann, J. E., D. E. Hudgens, J. Herter, and J. Martinich, 2010: Assessing sea-level rise impacts: A GIS-based framework and application to coastal New Jersey. *Coastal Management*, **38**, 433-455, doi:10.1080/08920753.2010.496105.

18. Biging, G., J. Radke, and J. H. Lee, 2012: Vulnerability assessments of transportation infrastructure under potential inundation due to sea-level rise and extreme storm events in the San Francisco Bay Region. *Paper for the California Vulnerability and Adaptation Study. Public Interest Energy Research Program. California Energy Commission Report,* **in press**.

19. AWF/AEC/Entergy, 2010: Building a Resilient Energy Gulf Coast: Executive Report, 11 pp., America's Wetland Foundation, America's Energy Coast, and Entergy. [Available online at www.entergy.com/content/our_community/environment/GulfCoastAdaptation/Building_a_Resilient_Gulf_Coast.pdf]

20. State of Louisiana, 2012: Louisiana's Comprehensive Master Plan for a Sustainable Coast, draft Jan 2012, State of Louisiana. Coastal Protection and Restoration Authority, Baton Rouge, LA. [Available online at http://www.coastalmasterplan.louisiana.gov/2012-master-plan/final-master-plan/]

21. NOAA, 1998: National Ocean Report. NOAA's Office of Public and Constituent Affairs. [Available online at http://www.publicaffairs.noaa.gov/oceanreport/tourism.html]

    U.S. Travel Association, cited 2012: U.S. Travel Forecasts. U.S Travel Association. [Available online at http://www.ustravel.org/sites/default/files/page/2009/09/ForecastSummary.pdf]

22. DOT, cited 2010: Freight Analysis Framework (Version 3) Data Tabulation Tool, Total Flows. U.S. Department of Transportation. [Available online at http://faf.ornl.gov/fafweb/Extraction1.aspx]

23. Burton, C., and S. Cutter, 2008: Levee failures and social vulnerability in the Sacramento-San Joaquin Delta area, California. *Natural Hazards Review*, **9**, 136, doi:10.1061/(ASCE)1527-6988(2008)9:3(136).

    Cutter, S. L., and C. Finch, 2008: Temporal and spatial changes in social vulnerability to natural hazards. *Proceedings of the National Academy of Sciences*, **105**, 2301-2306, doi:10.1073/pnas.0710375105.

    Emrich, C. T., and S. L. Cutter, 2011: Social vulnerability to climate-sensitive hazards in the southern United States. *Weather, Climate, and Society*, **3**, 193-208, doi:10.1175/2011WCAS1092.1. [Available online at http://journals.ametsoc.org/doi/pdf/10.1175/2011WCAS1092.1]

    Oxfam America, 2009: Exposed: Social Vulnerability and Climate Change in the US Southeast, 24 pp., Oxfam America Inc., Boston, MA. [Available online at http://adapt.oxfamamerica.org/resources/Exposed_Report.pdf]

    Rygel, L., D. O'Sullivan, and B. Yarnal, 2006: A method for constructing a Social Vulnerability Index: An application to hurricane storm surges in a developed country. *Mitigation and Adaptation Strategies for Global Change*, **11**, 741-764, doi:10.1007/s11027-006-0265-6. [Available online at http://www.cara.psu.edu/about/publications/Rygel_et_al_MASGC.pdf]]

24. Martinich, J., J. Neumann, L. Ludwig, and L. Jantarasami, 2013: Risks of sea level rise to disadvantaged communities in the United States. *Mitigation and Adaptation Strategies for Global Change*, **18**, 169-185, doi:10.1007/s11027-011-9356-0. [Available online at http://link.springer.com/content/pdf/10.1007%2Fs11027-011-9356-0]

25. Bovbjerg, R. R., and J. Hadley, 2007: Why Health Insurance Is Important. Report No. DC-SPG no. 1, 3 pp., The Urban Institute. [Available online at http://www.urban.org/UploadedPDF/411569_importance_of_insurance.pdf]

    Clark, G. E., S. C. Moser, S. J. Ratick, K. Dow, W. B. Meyer, S. Emani, W. Jin, J. X. Kasperson, R. E. Kasperson, and H. E. Schwarz, 1998: Assessing the vulnerability of coastal communities to extreme storms: The case of Revere, MA., USA. *Mitigation and Adaptation Strategies for Global Change*, **3**, 59-82, doi:10.1023/A:1009609710795.

Cutter, S. L., B. J. Boruff, and W. L. Shirley, 2003: Social vulnerability to environmental hazards. *Social Science Quarterly*, **84**, 242-261, doi:10.1111/1540-6237.8402002.

Moser, S. C., R. E. Kasperson, G. Yohe, and J. Agyeman, 2008: Adaptation to climate change in the Northeast United States: opportunities, processes, constraints. *Mitigation and Adaptation Strategies for Global Change*, **13**, 643-659, doi:10.1007/s11027-007-9132-3. [Available online at http://www.northeastclimateimpacts.org/pdf/miti/moser_et_al.pdf]

Texas Health Institute, 2012: Climate Change, Environmental Challenges and Vulnerable Communities: Assessing Legacies of the Past, Building Opportunities for the Future, 109 pp., The Joint Center for Political and Economic Studies Research Project, Washington, D.C. [Available online at http://www.jointcenter.org/docs/Climate_Change_Full_Report.pdf]

26. Cooley, H., E. Moore, M. Heberger, and L. Allen, 2012: Social Vulnerability to Climate Change in California. California Energy Commission. Publication Number: CEC-500-2012-013, 69 pp., Pacific Institute, Oakland, CA.

Bronen, R., 2011: Climate-induced community relocations: Creating an adaptive governance framework based in human rights doctrine. *NYU Review Law & Social Change*, **35**, 357-408. [Available online at http://socialchangenyu.files.wordpress.com/2012/08/climate-induced-migration-bronen-35-2.pdf]

Maldonado, J. K., C. Shearer, R. Bronen, K. Peterson, and H. Lazrus, 2013: The impact of climate change on tribal communities in the US: Displacement, relocation, and human rights. *Climatic Change*, **120**, 601-614, doi:10.1007/s10584-013-0746-z.

Titus, J. G., D. E. Hudgens, D. L. Trescott, M. Craghan, W. H. Nuckols, C. H. Hershner, J. M. Kassakian, C. J. Linn, P. G. Merritt, T. M. McCue, J. F. O'Connell, J. Tanski, and J. Wang, 2009: State and local governments plan for development of most land vulnerable to rising sea level along the US Atlantic coast. *Environmental Research Letters*, **4**, doi:10.1088/1748-9326/4/4/044008.

27. Holzman, D. C., 2012: Accounting for nature's benefits: The dollar value of ecosystem services. *Environmental Health Perspectives*, **120**, a152-a157, doi:10.1289/ehp.120-a152. [Available online at http://www.ncbi.nlm.nih.gov/pmc/articles/PMC3339477/pdf/ehp.120-a152.pdf]

Millennium Ecosystem Assessment, 2005: *Ecosystems and Human Well-Being. Health Synthesis*. Island press 53 pp.

Ruckelshaus, M., S. C. Doney, H. M. Galindo, J. P. Barry, F. Chan, J. E. Duffy, C. A. English, S. D. Gaines, J. M. Grebmeier, A. B. Hollowed, N. Knowlton, J. Polovina, N. N. Rabalais, W. J. Sydeman, and L. D. Talley, 2013: Securing ocean benefits for society in the face of climate change. *Marine Policy*, **40**, 154-159, doi:10.1016/j.marpol.2013.01.009.

Costanza, R., O. Pérez-Maqueo, M. L. Martinez, P. Sutton, S. J. Anderson, and K. Mulder, 2008: The value of coastal wetlands for hurricane protection. *AMBIO: A Journal of the Human Environment*, **37**, 241-248, doi:10.1579/0044-7447(2008)37[241:tvocwf]2.0.co;2. [Available online at http://www.bioone.org/doi/pdf/10.1579/0044-7447%282008%2937%5B241%3ATVOCWF%5D2.0.CO%3B2]

28. Principe, P., P. Bradley, S. H. Yee, W. S. Fisher, E. D. Johnson, P. Allen, and D. E. Campbell, 2012: Quantifying Coral Reef Ecosystem Services. EPA/600/R-11/206, 158 pp., U.S. Environmental Protection Agency, Office of Research and Development, Washington, D.C. [Available online at http://cfpub.epa.gov/si/si_public_record_report.cfm?dirEntryId=239984]

29. Hoegh-Guldberg, O., P. J. Mumby, A. J. Hooten, R. S. Steneck, P. Greenfield, E. Gomez, C. D. Harvell, P. F. Sale, A. J. Edwards, K. Caldeira, N. Knowlton, C. M. Eakin, R. Iglesias-Prieto, N. Muthiga, R. H. Bradbury, A. Dubi, and M. E. Hatziolos, 2007: Coral reefs under rapid climate change and ocean acidification. *Science*, **318**, 1737-1742, doi:10.1126/science.1152509.

30. Petes, L. E., A. J. Brown, and C. R. Knight, 2012: Impacts of upstream drought and water withdrawals on the health and survival of downstream estuarine oyster populations. *Ecology and Evolution*, **2**, 1712-1724, doi:10.1002/ece3.291. [Available online at http://onlinelibrary.wiley.com/doi/10.1002/ece3.291/pdf]

31. Barton, A., B. Hales, G. G. Waldbusser, C. Langdon, and R. A. Feely, 2012: The Pacific oyster, *Crassostrea gigas*, shows negative correlation to naturally elevated carbon dioxide levels: Implications for near-term ocean acidification effects. *Limnology and Oceanography*, **57**, 698-710, doi:10.4319/lo.2012.57.3.0698.

32. Hoegh-Guldberg, O., and J. F. Bruno, 2010: The impact of climate change on the world's marine ecosystems. *Science*, **328**, 1523-1528, doi:10.1126/science.1189930.

33. Goidel, K., C. Kenny, M. Climek, M. Means, L. Swann, T. Sempier, and M. Schneider, 2012: 2012 Gulf Coast Climate Change Survey Executive Summary. MASGP-12-017, 36 pp., National Oceanic and Atmospheric Administration, Texas Sea Grant, Louisiana Sea Grant, Florida Sea Grant, Mississippi-Alabama Sea Grant Consortium. [Available online at http://www.southernclimate.org/documents/resources/Climate_change_perception_survey_summary_NOAA_Sea_Grant_2012.pdf]

Responsive Management, 2010: Responsive Management: Delaware Residents' Opinions on Climate Change and Sea Level Rise, 351 pp., Responsive Management, Harrisonburg, VA. [Available online at http://www.dnrec.delaware.gov/coastal/Documents/SeaLevelRise/SLRSurveyReport.pdf]

Krosnik, J., 2013: Stanford University Climate Adaptation National Poll, 20 pp., Stanford Woods Institute for the Environment. [Available online at http://woods.stanford.edu/research/public-opinion-research/2013-Stanford-Poll-Climate-Adaptation]

34. Abbott, T., 2013: Shifting shorelines and political winds – The complexities of implementing the simple idea of shoreline setbacks for oceanfront developments in Maui, Hawaii. *Ocean & Coastal Management*, **73**, 13-21, doi:10.1016/j.ocecoaman.2012.12.010. [Available online at http://www.sciencedirect.com/science/article/pii/S0964569112003353]

Agyeman, J., P. Devine-Wright, and J. Prange, 2009: Close to the edge, down by the river? Joining up managed retreat and place attachment in a climate changed world. *Environment and Planning A*, **41**, 509-513, doi:10.1068/a41301. [Available online at http://www.envplan.com/epa/editorials/a41301.pdf]

Peach, S., 2012: Sea Level Rise, One More Frontier For Climate Dialogue Controversy. *Yale Forum on Climate Change and the Media*. Yale University. [Available online at http://www.yaleclimatemediaforum.org/2012/02/sea-level-rise-one-more-frontier-for-climate-dialogue-controversy/]

Kick, E. L., J. C. Fraser, G. M. Fulkerson, L. A. McKinney, and D. H. De Vries, 2011: Repetitive flood victims and acceptance of FEMA mitigation offers: An analysis with community–system policy implications. *Disasters*, **35**, 510-539, doi:10.1111/j.1467-7717.2011.01226.x.

35. Rosenzweig, C., W. D. Solecki, R. Blake, M. Bowman, C. Faris, V. Gornitz, R. Horton, K. Jacob, A. LeBlanc, R. Leichenko, M. Linkin, D. Major, M. O'Grady, L. Patrick, E. Sussman, G. Yohe, and R. Zimmerman, 2011: Developing coastal adaptation to climate change in the New York City infrastructure-shed: Process, approach, tools, and strategies. *Climatic Change*, **106**, 93-127, doi:10.1007/s10584-010-0002-8. [Available online at http://www.ccrun.org/sites/ccrun/files/attached_files/2011_Rosenzweig_etal.pdf]

Hudson, B., 2012: Coastal land loss and the mitigation – adaptation dilemma: Between Scylla and Charybdis. *Louisiana Law Review*, **73**. [Available online at http://digitalcommons.law.lsu.edu/lalrev/vol73/iss1/3]

IPCC, 2012: *Managing the Risks of Extreme Events and Disasters to Advance Climate Change Adaptation. A Special Report of Working Groups I and II of the Intergovernmental Panel on Climate Change.* C. B. Field, V. Barros, T.F. Stocker, D. Qin, D. J. Dokken, K. L. Ebi, M. D. Mastrandrea, K. J. Mach, G.-K. Plattner, S.K. Allen, M. Tignor, and P. M. Midgley, Eds. Cambridge University Press, 582 pp. [Available online at http://ipcc-wg2.gov/SREX/images/uploads/SREX-All_FINAL.pdf]

36. Multihazard Mitigation Council, 2005: Natural Hazard Mitigation Saves: An Independent Study to Assess the Future Savings From Mitigation Activities. Volume 2 - Study Documentation, 150 pp., National Institute of Building Sciences, Washington, D.C. [Available online at http://www.nibs.org/resource/resmgr/MMC/hms_vol2_ch1-7.pdf?hhSearchTerms=Natural+and+hazard+and+mitigation]

37. Cropper, M. L., and P. R. Portney, 1990: Discounting and the evaluation of life-saving programs. *Journal of Risk and Uncertainty*, **3**, 369-379, doi:10.1007/BF00353347.

38. Franck, T., 2009: Coastal adaptation and economic tipping points. *Management of Environmental Quality: An International Journal*, **20**, 434-450, doi:10.1108/14777830910963762.

Hallegatte, S., 2012: A framework to investigate the economic growth impact of sea level rise. *Environmental Research Letters*, **7**, 015604, doi:10.1088/1748-9326/7/1/015604.

H. John Heinz III Center for Science Energy and the Environment, 2000: *The Hidden Costs of Coastal Hazards: Implications for Risk Assessment and Mitigation.* A multisector collaborative project of the H. John Heinz Center for Science, Economics, and the Environment. Island Press, 252 pp.

39. Barthel, F., and E. Neumayer, 2010: A trend analysis of normalized insured damage from natural disasters. *Climatic Change*, **113**, 215-237, doi:10.1007/s10584-011-0331-2.

40. Burkett, V., and M. Davidson, 2012: *Coastal Impacts, Adaptation and Vulnerabilities: A Technical Input to the 2013 National Climate Assessment.* Island Press, 216 pp.

41. GAO, 2010: National Flood Insurance Program: Continued Actions Needed to Address Financial and Operational Issues. U.S. Government Accountability Office, Washington, D.C. [Available online at http://www.gao.gov/assets/130/124468.pdf]

Ntelekos, A. A., M. Oppenheimer, J. A. Smith, and A. J. Miller, 2010: Urbanization, climate change and flood policy in the United States. *Climatic Change*, **103**, 597-616, doi:10.1007/s10584-009-9789-6.

42. FEMA, cited 2013: Total Coverage by Calendar Year. U.S. Federal Emergency Management Agency. [Available online at http://www.fema.gov/policy-claim-statistics-flood-insurance/policy-claim-statistics-flood-insurance/policy-claim-13-12]

43. H. John Heinz III Center for Science Energy and the Environment, 2000: Evaluation of Erosion Hazards, 205 pp, Washington, DC, USA. [Available online at http://www.fema.gov/pdf/library/erosion.pdf]

Czajkowski, J., H. Kunreuther, and E. Michel-Kerjan, 2011: A Methodological Approach for Pricing Flood Insurance and Evaluating Loss Reduction Measures: Application to Texas, Wharton Risk Management Center and CoreLogic, Philadelphia, PA and Santa Ana, CA, 87 pp. [Available online at http://opim.wharton.upenn.edu/risk/library/WhartonRiskCenter_TexasFloodInsurancePricingStudy.pdf]

Kunreuther, H. C., and E. O. Michel-Kerjan, 2009: *At War with the Weather: Managing Large-Scale Risks in a New Era of Catastrophes.* The MIT Press, 416 pp.

Michel-Kerjan, E., and H. Kunreuther, 2011: Redesigning flood insurance. *Science*, **333**, 408-409, doi:10.1126/science.1202616. [Available online at http://erwannmichelkerjan.com/wp-content/uploads/2011/07/RedesigningFloodIns_ScienceMag_20110722-1.pdf]

Michel-Kerjan, E. O., 2010: Catastrophe economics: The National Flood Insurance Program. *The Journal of Economic Perspectives*, **24**, 165-186, doi:10.1257/jep.24.4.165. [Available online at http://www.jstor.org/stable/pdfplus/20799178.pdf]

King, R. O., 2011: National Flood Insurance Program: Background, Challenges, and Financial Status. R40650, 33 pp., Congressional Research Service, Washington, D.C. [Available online at http://www.fas.org/sgp/crs/misc/R40650.pdf]

44. Colls, A., N. Ash, and N. Ikkala, 2009: *Ecosystem-based Adaptation: A Natural Response to Climate Change.* International Union for Conservation of Nature and Natural Resources, 16 pp. [Available online at http://data.iucn.org/dbtw-wpd/edocs/2009-049.pdf]

Danielsen, F., M. K. Sørensen, M. F. Olwig, V. Selvam, F. Parish, N. D. Burgess, T. Hiraishi, V. M. Karunagaran, M. S. Rasmussen, L. B. Hansen, A. Quarto, and N. Suryadiputra, 2005: The Asian tsunami: A protective role for coastal vegetation. *Science*, **310**, 643, doi:10.1126/science.1118387.

Swann, L. D., 2008: The Use of Living Shorelines to Mitigate the Effects of Storm Events on Dauphin Island, Alabama, USA. American Fisheries Society Symposium 12 pp., Department of Fisheries and Allied Aquaculture, Auburn University, Ocean Springs, MS. [Available online at http://livingshorelinesolutions.com/uploads/Dr._LaDon_Swann__Living_Shorelines_Paper.pdf]

The World Bank, 2009: *Convenient Solutions for an Inconvenient Truth: Ecosystem-based Approaches to Climate Change.* The World Bank, The International Bank for Reconstruction and Development, 91 pp.

Tobey, J., P. Rubinoff, D. Robadue Jr, G. Ricci, R. Volk, J. Furlow, and G. Anderson, 2010: Practicing coastal adaptation to climate change: Lessons from integrated coastal management. *Coastal Management*, **38**, 317-335, doi:10.1080/08920753.2010.483169.

UNEP-WCMC, 2006: *In the Front Line: Shoreline Protection and Other Ecosystem Services From Mangroves and Coral Reefs.* UNEP-WCMC, 33 pp. [Available online at http://www.unep.org/pdf/infrontline_06.pdf]

Villanoy, C., L. David, O. Cabrera, M. Atrigenio, F. Siringan, P. Aliño, and M. Villaluz, 2012: Coral reef ecosystems protect shore from high-energy waves under climate change scenarios. *Climatic Change*, **112**, 1-13, doi:10.1007/s10584-012-0399-3.

45. Daily, G. C., S. Polasky, J. Goldstein, P. M. Kareiva, H. A. Mooney, L. Pejchar, T. H. Ricketts, J. Salzman, and R. Shallenberger, 2009: Ecosystem services in decision making: Time to deliver. *Frontiers in Ecology and the Environment*, **7**, 21-28, doi:10.1890/080025.

Koch, E. W., E. B. Barbier, B. R. Silliman, D. J. Reed, G. M. E. Perillo, S. D. Hacker, E. F. Granek, J. H. Primavera, N. Muthiga, S. Polasky, B. S. Halpern, C. J. Kennedy, C. V. Kappel, and E. Wolanski, 2009: Non-linearity in ecosystem services: Temporal and spatial variability in coastal protection. *Frontiers in Ecology and the Environment*, **7**, 29-37, doi:10.1890/080126. [Available online at http://www.esajournals.org/doi/pdf/10.1890/080126]

46. Georgetown Climate Center, cited 2012: Helping Communities Adapt to Climate Change. [Available online at http://www.georgetownclimate.org/adaptation/overview]

Moser, S. C., and J. A. Ekstrom, 2012: Identifying and Overcoming Barriers to Climate Change Adaptation in San Francisco Bay: Results from Case Studies. Publication number: CEC-500-2012-034 186 pp., California Energy Commission, Sacramento, CA. [Available online at http://www.energy.ca.gov/2012publications/CEC-500-2012-034/CEC-500-2012-034.pdf]

NPCC, 2010: *Climate Change Adaptation in New York City: Building a Risk Management Response: New York City Panel on Climate Change 2009 Report.* Vol. 1196, C. Rosenzweig, and W. Solecki, Eds. Wiley-Blackwell, 328 pp. [Available online at http://onlinelibrary.wiley.com/doi/10.1111/nyas.2010.1196.issue-1/issuetoc]

——, 2009: Climate Risk Information, 74 pp., New York City Panel on Climate Change, New York, New York. [Available online at http://www.nyc.gov/html/om/pdf/2009/NPCC_CRI.pdf]

47. Schmidtlein, M. C., R. C. Deutsch, W. W. Piegorsch, and S. L. Cutter, 2008: A sensitivity analysis of the social vulnerability index. *Risk Analysis*, **28**, 1099-1114, doi:10.1111/j.1539-6924.2008.01072.x. [Available online at https://groups.nceas.ucsb.edu/sustainability-science/2010%20weekly-sessions/session-3-09.27.2010/supplemental-readings-from-princeton-group/misc-ideas-papers/Schmidtlein%20et%202008%20sensitiv%20analysis%20of%20vuln%20Indiex.pdf]

48. Regional Threats from Climate Change are compiled from technical input reports, the regional chapters in this report, and from scientific literature.

Future National Assessments

1. Cash, D. W., and S. C. Moser, 2000: Linking global and local scales: Designing dynamic assessment and management processes. *Global Environmental Change*, **10**, 109-120, doi:10.1016/S0959-3780(00)00017-0.

Clark, W. C., R. B. Mitchell, and D. W. Cash, 2006: Ch. 1: Evaluating the influence of global environmental assessments. *Global Environmental Assessments: Information and Influence*, R. B. Mitchell, W. C. Clark, D. W. Cash, and N. Dickson, Eds., The MIT Press, 1-26.

Farrell, A., and J. Jäger, Eds., 2005: *Assessments of Regional and Global Environmental Risks: Designing Processes for the Effective Use of Science in Decision-Making.* Resources for the Future, 301 pp. [Available online at http://www.amazon.com/Assessments-Regional-Global-Environmental-Risks/dp/1933115041]

Mitchell, R. B., W. C. Clark, D. W. Cash, and N. M. Dickson, Eds., 2006: *Global Environmental Assessments: Information and Influence.* MIT Press, 352 pp.

NRC, 2007: *Analysis of Global Change Assessments: Lessons Learned.* National Research Council, Committee on Analysis of Global Change Assessments, Board on Atmospheric Sciences and Climate, Division on Earth and Life Studies. National Academies Press, 196 pp. [Available online at http://www.nap.edu/catalog.php?record_id=11868]

2. Buizer, J., P. Fleming, S. L. Hays, K. Dow, C. Field, D. Gustafson, A. Luers, and R. H. Moss, 2013: Preparing the Nation for Change: Building a Sustained National Climate Assessment. National Climate Assessment and Development Advisory Committee, Washington, D.C. [Available online at http://www.nesdis.noaa.gov/NCADAC/pdf/NCA-SASRWG%20Report.pdf]